The Natural History of the Gorilla

The Natural History
of the Gorilla

A.F. Dixson

With a foreword by R.D. Martin

New York Columbia University Press 1981

Photography by T. Dennett
Wellcome Laboratories of Comparative Physiology
Zoological Society of London

Printed in Great Britain

Library of Congress Cataloging in Publication Data

Dixson, A F
    The natural history of the gorilla.
    Bibliography: p.
    Includes index.
    1. Gorillas. I. Title.
QL737.P96D59   599.88'46   81–57
ISBN 0–231'–05318'–5

*'. . . he is more like a giant in stature than a man;
for he is very tall, and hath a man's face,
hollow-eyed, with longe haire upon his browes'*

Andrew Battell (1625)

Contents

Plates

PLATES

Figures

FIGURES

Tables

Preface

Despite the great interest which the gorilla holds for scientist and layman alike, no detailed review has been attempted since Schaller's *Mountain Gorilla: Ecology and Behavior* of 1963. The present volume is a 'Natural History' and it is not intended to be a textbook. However, I have tried to provide an up-to-date review of many aspects of the gorilla's biology which may interest students of zoology, anthropology or psychology and not only those who specialize in the study of primates. I hope that the book may be readable as well as informative. References are not cited in the text as in a more formal treatise but the authors whose researches are described are named in the text and a complete chapter-by-chapter bibliography has been provided. Those who wish to pursue some problem in greater depth should be able to trace the relevant sources.

Many people have allowed me to reproduce data from their published and unpublished work. Their names are cited in the captions to the tables, figures and plates and I am extremely grateful to all concerned. I should also like to thank the following organizations for permission to present material (tables, figures or plates) either in the original or in modified form:

Academic Press Inc. (Fig. 3D and G; Fig. 26; Table 3); American Anthropological Association (Table 13); Anthropological Institute of Great Britain and Ireland (Fig. 9); Cambridge University Press (Figs 27, 37; Table 9); Chicago University Press (Figs 15, 20, 22); the editor and publishers of *Genetic Psychology Monographs* (Plate 8); Edinburgh University Press (Figs 2, 3A, B and C); Holt, Saunders Ltd (chimpanzee data in Fig. 41; Tables 4 and 5); Fauna Preservation Society (Plate 18; Table 16); Japan Science Press (Table 14); Jersey Wildlife Preservation Trust (Figs 28A, 49, 50); Karger Ltd (Fig.33; Plate 9 and 15); Macmillan Ltd (Figs 8, 31, 32); Marquis and Co. Ltd (Plate 1); Plenum Publishing Co. (Figs 35, 36); Rosenburg and Sellier Ltd (Plate 7); Survival Anglia Ltd (Plate 17); Verlag Paul Parey (Figs 38, 39); Weidenfeld and Nicolson Ltd (Figs 6B, 48); Yale University Press (Fig. 34) and the Zoological Society of London (Fig. 16; Plates 3, 4, 5A, 6, 19, 21; Table 11).

The National Geographic Society and Fauna Preservation Society have financed several expeditions to 'census' and study the mountain gorillas of the Virunga volcanoes. I took part in the final census in 1973. I am very grateful to have been given this opportunity to observe mountain gorillas in the wild, and in particular to Dr Dian Fossey who supervised the field work. Dr Graeme Groom played a key role in all the census expeditions and it is a pleasure to thank him for his advice during the writing of this book. Originally we planned it as a joint venture, but his medical duties prevented him from taking part. His suggestions for improving the final chapter were invaluable. Mrs P. Napier of the British Museum of Natural History kindly allowed me access to the museum's collection of skeletal material from gorillas. David Jones and the staff of the Pathology Department at the Zoological Society of London were most helpful in providing tissues from dead gorillas, orang-utans and chimpanzees. Some material was also provided by Chester Zoo, Bristol Zoo and Yerkes Primate Research Centre. I also wish to thank the staff of the Zoological Society's Library and the Librarian at Oregon Regional Primate Research Centre who obtained a great many works of reference for me when I was writing earlier drafts. Mr Joel Ito of the Oregon Primate Centre has kindly allowed me to use several of his drawings in the book; I wish there were more, since my own illustrations lack his professional skill. Many of the graphs were drawn by Mr D. Fleming, and the manuscript was typed, and retyped, without complaint by Miss C. Nutkins, Miss A. Cannon, Mrs J. Hill, Mrs R. Simper and my wife Amanda. My thanks to them all.

Dr A.H. Harcourt and Dr R.D. Martin have read and criticized portions of the manuscript and this was most helpful. Any errors which occur are, of course, my own responsibility. I am grateful to the Mental Health Foundation (Great Britain) and the Wellcome Trust who supported me as a post-doctoral fellow during much of the time that this book was written.

Dr Alan F. Dixson
Zoological Society of London
April 1980

Foreword

Over a century has elapsed since Charles Darwin and Thomas Henry Huxley demonstrated that the great apes – chimpanzees, gorillas and orang-utans – are our closest zoological relatives. It is now widely accepted that the African chimpanzees and gorillas are closer to human ancestry than the Asiatic orangs, and that a common ancestor of man and the African apes existed at some time between six and twenty million years ago. Given that we have had over a hundred years for research, it is surprising that we are still so vague about the time of the divergence between man, chimpanzee and gorilla. We do not even have reliable evidence as to whether it is the chimpanzee or the gorilla which is closest to us in an evolutionary sense. Despite many claims to the contrary – largely fostered by those who study chimpanzees – it has not been scientifically established that the gorilla is the less closely related to human origins. A fair assessment, given the present state of our knowledge, is that the divergence between man, chimp and gorilla should be represented as a three-way split with dotted lines. In short, it is quite possible that further research will demonstrate that the gorilla is our closest relative in the grand evolutionary tree of the animal kingdom. In any event, it comes very close indeed and is of particular interest for this reason. It is therefore shocking that we know so little about the gorilla, and it is a considerable pleasure for me to be able to introduce this thorough and readable book on the subject by Alan Dixson.

It is an unfortunate fact that many zoologists today tend to specialize so that they study animals either in the field or in the laboratory. In fact, many zoologists concerned in laboratory studies – including some conducted on the gorilla – may never see the animal itself but only some fragment, such as a blood sample. This is a pity because it is really necessary to combine both field and laboratory studies in order to reach a genuine scientific understanding. Field studies provide a unique framework for studying an animal in the natural environment to which it is adapted both structurally and behaviourally. Laboratory studies, on the other hand, permit more detailed investigation of the anatomical, physiological and behavioural components which are integrated in the wonderfully adapted whole. It is for this reason that Dr Dixson is particularly well qualified as the author of this book. He is an able laboratory investigator who has made time in the course of his research work to take part in field studies, including a valuable census of mountain gorillas. This combination of expertise has produced a finely balanced treatment of the gorilla, ranging from its basic anatomy to its present status in the wild.

The question of conservation, examined in detail in the last chapter, is of particular importance. It is, in a way, a revealing comment on the current conservation measures in Africa that the Virunga Park of Zaire (which was the first National Park to be set up in Africa, fifty-three years ago) includes one of the largest populations of the severely endangered mountain gorilla subspecies of which something less than 300 now survive. It is clear from this that without the protective measures taken by the Government of Zaire the plight of the mountain gorillas would be even worse but we have to face the possibility that mountain gorillas will become extinct within the next few years. When a population is reduced to less than 300 individuals, it is likely that genetic variability has been reduced beyond the minimum required for survival. The scanty evidence available indicates that the status of lowland gorillas may be somewhat better, but even there the outlook is bleak. We may soon be in a situation where one of our closest zoological relatives has disappeared from the face of the earth, barely a century after the discovery of how much it had to teach us about human evolution. Yet even this scientific loss seems pale beside the major tragedy, not only of losing yet another animal species, but also of the wholesale destruction of the African rain-forest and *all* of the animal and plant species which live in this complex network of interrelationships. One can only hope that this informative book will play its part in forestalling this catastrophe.

R.D. Martin
University College London

Chapter 1

Historical Perspective

OF all the types of animal I have observed naturalistically or used in experiments during almost 60 years as a psychobiologist, none has stirred my curiosity and suggested so many questions as the gorilla.

When Robert Yerkes wrote the above words, almost thirty years ago, he also expressed concern at the lack of information on all aspects of the gorilla's biology and pointed out the need for conservation schemes to prevent its extermination. Today, much more is known about gorillas, particularly about their behaviour and ecology, but deforestation, hunting and collecting have greatly reduced their numbers.

There are several reasons why science has been slow to investigate a creature which is one of man's closest relatives. The gorilla has a limited distribution range within equatorial Africa and, during recent times at least, has probably never been common. It inhabits dense forests where it is difficult to observe or capture. Moreover, even nowadays it is no easy matter to keep gorillas alive and healthy in captivity; only during the last twenty-five years have they reproduced successfully in zoos. Adult gorillas are extremely powerful animals and the chimpanzee has usually proved a more practical choice for experimental studies.

Of the five genera of apes, the gorilla was the last to be discovered for science, no specimens being obtained until 1847. The gibbon, siamang and orang-utan of Southeast Asia all appear in zoological treatises long before this. Buffon observed a live gibbon and wrote an account of it in the fourteenth volume of his *Histoire Naturelle*, published in 1766. Sir Stamford Raffles described the siamang in 1822. A description of the orang-utan was provided in 1658 by Jacob Bontius in his *Historiae Naturalae & Medicae Indiae Orientalis*. One illustration in the book has been aptly described by T.H. Huxley as resembling 'nothing but a very hairy woman of rather comely aspect'. More accurate descriptions of the orang-utan were written by

Vosmaer in 1778 and Peter Camper in 1779. Nevertheless, confusion over the number of sorts of orang-utan continued for many years because specimens initially brought from Borneo to Europe were all of immature animals. When adult males were eventually obtained, their markedly different appearance led to their classification as a new species, the 'Pongo'.

Turning to African apes, the chimpanzee was discovered and its anatomy extensively investigated long before the gorilla was proved to exist. Nicolaas Tulp, of Amsterdam, gave an account of a chimpanzee, or *Satyrus indicus* as he called it, in 1641. The illustration in his book looks like an imaginative and lugubrious blend of chimpanzee and orang-utan, but it is recognizable as an example of the former. Vernon Reynolds has even suggested that it may have been an example of the rare pigmy chimpanzee (*Pan paniscus*) rather than of *P. troglodytes* as is usually thought to be the case. In 1699 a treatise by Edward Tyson appeared, entitled *Orangutang, sive Home sylvestris; or the anatomy of a pigmie compared with that of a monkey, an ape and a man*. The 'pigmie', actually a young chimpanzee, is shown standing upright and supporting itself with a cane, the usual practice in an age which regarded apes as quasi-human creations. Tyson's beautifully illustrated account of the anatomy of the chimpanzee is still an acknowledged masterpiece.

Long before scientists were able to examine the chimpanzee, however, it must either have been sighted by explorers visiting western Africa or was described to them by the native inhabitants. One such adventurer, a Portuguese named Edouardo Lopez, recounts having seen many 'apes' during his travels in Africa and, in 1598, Pigafetta included an illustration of these animals in his *Report of the Kingdom of the Congo*. This depicts long-armed, tail-less creatures which might be chimpanzees or gorillas, but cannot be reliably identified. A much more definite indication of the occurrence of apes in Africa is to be found in *Purchas his Pilgrims*, published in 1625 and including a description by Andrew Battell (a seaman from Leigh in Essex) of 'two kinds of monsters which are common in these woods and very dangerous. The greater of these two monsters is called Pongo in their language and the lesser is called Engeco.' In all probability the Engeco was the chimpanzee whilst the Pongo, not to be confused with the orang or Pongo of Southeast Asia, was the gorilla. Battell's account has often been quoted but it is such a charming mixture of reality and nonsense that it will perhaps bear further repetition.

This Pongo is in all proportion like a man; but that he is more like a giant in stature than a man; for he is very tall, and hath a man's face, hollow-eyed, with long haire upon his browes. His face and eares are without haire, and his hands also. His bodie is full of haire, but not very thicke; and it is of a dunnish colour.

He differeth from a man but in his legs; for they have no calfe. Hee goeth alwaies upon his legs and carrieth his hands clasped in the nape of his necke when he goeth upon the ground. They sleepe in the trees and build shelters for the raine. They feed upon fruit that they find in the woods, and upon nuts, for they eate no kind of flesh. They cannot speake, and have no understanding more than a beast. The people of the countrie, when they travaile in the woods make fires where they sleepe in the night; and in the morning, when they are gone, the Pongoes will come and sit about the fire till it goeth out; for they have no understanding to lay the wood together. They goe many together and kill many negroes that travaile in the woods. Many times they fall upon the elephants which come to feed where they be, and so beate them with their clubbed fists, and pieces of wood, that they will runne roaring away from them. Those Pongoes are never taken alive because they are so strong, that ten men cannot hold one of them, but yet they take many of their young ones with poisoned arrows.

The young Pongoe hangeth on his mother's bellie with his hands clasped about her, so that when the countrie people kill any of the females they take the young one, which hangeth fast upon his mother. When they die among themselves, they cover the dead with great heaps of boughs and wood, which is commonly found in the forest.

After Andrew Battell's tale of the Pongo, almost two centuries passed before another traveller returned from Africa to publish tales of a giant, man-like ape. In 1819 Thomas Bowdwich wrote a book entitled *Mission from Cape Coast to Ashantee* in which he described the 'Ingena', a beast 'five feet tall and four across the shoulders'. This, he informs us, 'is commonly seen by those who travel to Kaylee, lurking in the bush to destroy passengers, and feeding principally upon wild honey which abounds'. Bowdwich also says that his Ingena had the habit of building 'a house in rude imitation of the natives and sleeping outside or on the roof of it'. Actually, gorillas, like chimpanzees and orang-utans, do build nests each night and sleep on top of them.

Specimens of the gorilla were first obtained for scientific study in 1847, by an American medical missionary named Thomas Savage who was visiting what is now called Gabon. Savage relates:

Soon after my arrival Mr Wilson showed me a skull, represented by the natives to be that of a monkey-like animal, remarkable for its size, ferocity and habits.

From the contour of the skull and the information derived from several intelligent natives I was induced to believe that it belonged to a new species of orang.

The term 'orang' was then applied to any of the larger apes; the chimpanzee for instance was called the 'black orang' or *Troglodytes niger*. Savage was familiar with the chimpanzee for, several years previously, in association with Jeffries Wyman of Harvard University, he had written an account of its structure and habits. This is why he immediately recognized the gorilla skull as belonging to a new species for it had much more prominent brow bridges as well as bony crests (saggital and occipital crests) on top of the skull. He collected four skulls (two of each sex), a male and a female pelvis, some limb bones, vertebrae and ribs for shipment back to America. In 1847 Savage and Wyman wrote an account of the gorilla in the *Boston Journal of Natural History*; this was entitled *Notice of the external characters and habits of Troglodytes gorilla a new species of orang from the Gaboon River; osteology of the same*. The name 'gorilla has an ancient origin, deriving from an account written in the fifth century BC by Hanno, a Carthaginian who had sailed along the west coast of Africa and had killed 'hairy people' (the 'gorillae') which were probably some type of monkey.

The gorillas discovered by Thomas Savage belonged to the western lowland subspecies, now called *Gorilla g. gorilla*, which occurs in various parts of western central Africa. Eastern gorillas were not known until 1902 when Oscar von Beringe, a captain in the Belgian army, shot gorillas in the Virunga volcanoes and sent a specimen back to Europe. The mountain subspecies, *Gorilla g. beringei*, is named after him.

Knowledge of the gorilla's anatomy, particulary of its skeleton, was obtained long before accurate descriptions of its behaviour. Bones could survive the long sea journey back to Europe or America whereas living animals usually perished, and the prospect of studying what was thought to be a ferocious monster in its fever-ridden native jungles was not an enticing one. Savage relied on information provided by the Empongwe tribes-people concerning the gorilla's habits in the wild. These people, however, regarded the apes as 'degenerated human beings' or 'wild men of the woods'.

The scientific literature of the nineteenth and early twentieth centuries is filled with bizarre anecdotes about gorilla behaviour. In fairness to these earlier reviewers one must remember that they inhabited a mental climate in which many tribal societies were considered to be little better than animals, and in which the beasts of the jungle were thought to be constantly at war with one another. The gorilla became the centre of great interest when, in 1865, the eminent anatomist Richard Owen published his *Memoir on the Gorilla*, asserting that of all the apes the gorilla is closest structurally to man. This came shortly after the publication (in 1859) of Darwin's *Origin of Species*, at a time when new avenues of thought concerning human origins were opening up.

Perhaps because of his belief in the close kinship between gorilla and man, Owen was willing to accept and quote fantastic stories about its behaviour, such as the following:

If the old male be seen alone, or when in quest of food, he is usually armed with a stout stick, which the negroes aver to be the weapon with which he attacks his chief enemy the elephant. When therefore he discerns the elephant pulling down and wrenching off the branches of a favourite tree, the gorilla, stealing along the bough, strikes the sensitive proboscis of the elephant with a violent blow of his club.

When he tired of beating elephants, apparently the gorilla liked to battle with leopards (see Plate 1) or turned to homicide, and particularly enjoyed throttling the local inhabitants.

Negroes when stealing through the shades of the tropical forest become sometimes aware of the proximity of one of these frightfully formidable apes by the sudden disappearance of one of their companions, who is hoisted up into the tree, uttering, perhaps, a short choking cry. In a few minutes he falls to the ground a strangled corpse.

According to many narratives, chimpanzees and gorillas were also accomplished rapists. Bart, writing in 1833, recounts that chimps carried off young women 'who sometimes escaped to human society after having been for years detained by their ravishers in a frightful captivity'. Once the gorilla was discovered, he joined the chimpanzee in these nefarious activities and, in the guise of King Kong, he continues to the present day as an abductor of maidens. Some years ago a young gorilla named Achilles at Basle Zoo achieved notoriety after a female keeper had accidently locked herself in his cage. It was the following day before she was released, somewhat shaken but unhurt. The incident was reported in the press as an amorous adventure between ape and girl but, ironically, Achilles had been wrongly sexed by zoo authorities; 'he' turned out to be a female and later became the mother of the first gorilla born in Europe.

Oken once compared the behaviour of apes and man in the following terms: 'The apes resemble man in all bad moral traits; they are malicious, treacherous, thievish and indecent.' The truth of the matter is, of course, that man sometimes ascribes to apes and monkeys the traits he most despises in himself. Gorillas are harmless vegetarians and they avoid elephants, leopards and humans whenever possible. Their sexual activity, when viewed from our standpoint, is infrequent and fairly unimaginative.

When Rupert Garner attempted a field study of gorilla behaviour at the end of the last century, he had a large cage constructed which he sat inside lest the beasts of the forest should attack him. He saw few apes, but occasionally one or two passed near his cage.

One day as I sat alone, a young gorilla, perhaps five years old, came within six or seven yards of the cage and took a peep. He stood for a while almost erect, with one hand holding on to a bough, his lower lip was relaxed and the end of his tongue could be seen between his parted lips. He did not evince either fear or anger but rather appeared to be amazed.

Presumably white men crouching in cages were not a common sight in the rain forest. The tongue protrusion – accurately described by Garner – occurs in gorillas, as in man, when they are concentrating or are suprised and nervous. Garner was less accurate in his other descriptions of gorilla behaviour for, since he could not see the animals himself, he repeated as if they were factual many tales told to him by the natives. He tells us the gorilla has 'an incipient idea of government' as well as 'a faint perception of order and justice, if not of right and wrong'. Gorillas were believed to hold 'palavers' in the jungle during which the 'king' (adult male) sat in the centre and the others gathered round and talked excitely.

Garner did, however, dismiss as nonsense earlier tales about the ferocity of gorillas, many of which had been exaggerated by hunters in order to make their own exploits seem more courageous. The man most criticized in this regard was Paul Du Chaillu who, in 1861 and 1869, published books about his adventures in western central Africa. Actually he obtained useful information about the life and habits of gorillas but, perhaps at the request of his publishers, greatly embroidered his narratives and so eventually fell into disrepute. A reviewer of the time described Du Chaillu as 'a traveller who, forsaking all beaten tracks, plunged into the wilds of a country where no white man appears to have preceded him and who brings before us tribes marked by hideous moral degradation and yet not unhopeful aspects.' He described some very odd creatures indeed, including the 'Koola-Kamba' which, in the guise of a possible hybrid between the chimpanzee and gorilla, continued to stalk the pages of zoological texts until quite recently. He also shot the 'Nsiego Mbouvé', a bald-headed ape which built parasol-like structures in the trees and sheltered beneath them during rain-storms. This was probably a chimpanzee. Du Chaillu's descriptions of the dangers of gorilla hunting are very colourful. He refers to the adult male as 'an impossible piece of hideousness ... as aweful as a nightmare dream ... One blow of that huge paw, with its bony claws, and the poor hunter's entrails are torn out, his breast bone broken or his skull crushed.'

Probably it is a combination of the gorilla's formidable appearance and massive strength together with the impressive displays it employs to intimidate attackers which is responsible for its unfortunate reputation. A young female studied by Yerkes weighed only sixty-five pounds, yet with her feet firmly braced she could pull with her arms 160 pounds on a heavy spring balance. What an adult male weighing upwards of 300 pounds is capable of, we can only guess. 'Bushman', a very big male who lived for many years in Chicago's Lincoln Park Zoo, could burst footballs by squeezing them under his arm. In the wild, one of the most important social roles of such a male is to protect the females and young animals in his group. He may display at intruders by beating his chest, roaring loudly and tearing up vegetation. In dire circumstances he may charge at a hunter but almost invariably this is a bluff and the animal eventually turns and flees. These defensive tactics are extremely convincing and one might be pardoned for assuming that the gorilla means to attack. Natural selection has played the gorilla an ironic trick, however, for as he rises and beats his chest, the male presents an ideal target for a hunter's rifle.

Occasionally gorillas do attack hunters: the usual injuries are lacerations made by the animal's fingernails as it attempts to sweep its attacker aside and escape. Fred Merfield, a famous gorilla hunter, once had his thigh sliced open in this manner by a charging silverback (the adult male gorilla which leads the group). During a field study in 1959 two Japanese workers Kawai and Mizuhara, tracked a gorilla group too closely and were charged by its leader. A tracker was at the head of the party and 'the male first crashed into him and ran over him, crushing him under his heavy weight ... The other tracker stood still and sung [sic] his panga against the animal. The gorilla dashed at and beat him on the arm, and there ceased to attack and retraced his path.' Such incidents are extremely rare. Sabater Pi recorded all gorilla attacks on humans over ten years in the former Spanish colony of Rio Muni. Only seven attacks occurred, six of which were defensive measures against native hunters, whilst the remaining case involved a group of gorillas which panicked when it suddenly encountered a man on a narrow forest trail.

Many gorilla hunters have written books about their exploits but these tell us very little about the normal life of the animal. Merfield made some useful observations on western lowland gorillas and Carl Akeley became the first man to film gorillas when he visited the Virunga volcanoes in the 1920s. Much of our knowledge concerning

7

the geographical distribution of the gorilla, its anatomy, taxonomy, skeletal growth and many other aspects of its biology derives from studies of animals shot by hunters or collectors. A recent taxonomic study of the gorilla by Colin Groves was able to call upon 747 skulls of adults plus those of 'numerous juveniles' contained in thirty-five European and American museum collections. One of these was the Powell Cotton Museum, at Birchington in Kent. Some idea of Major Powell Cotton's prowess as a collector may be gleaned from the fact that his museum contains the remains of 217 gorillas. British collections alone were estimated in 1966 to contain 728 specimens of gorillas, all but twenty of which belonged to the western lowland subspecies.

The conclusion that the gorilla has been over-collected is certainly justified and its threatened extinction is at least partly due to this fact. Some collectors such as Carl Akeley and Henry Raven showed proper restraint in killing only the minimum of specimens necessary for their work. Akeley, moreover, was convinced that the mountain gorillas of the Virunga volcanoes should be protected and studied in their natural habitat. The Virungas are now a National Park but no extensive fieldwork took place there until 1959, when Emlen and Schaller began the first thorough study of gorilla ecology and behaviour.

Other collectors, less enlightened than Akeley and Raven, made a business of killing animals, as the following comment, made by William Gregory, in 1936, indicates:

In previous years collectors had organized drives and secured great numbers of gorillas' skins and skeletons. Moreover, Mr Raven had opportunity to witness the unfortunate effect (in Cameroun), as far as the protection of the gorilla was concerned, of the demand for gorilla skulls on the part of scientists, to such a degree that white men as well as natives had in the past done a profitable business in killing the animals.

Collectors also attempted to bring living specimens back to Europe or America. Ben Burbridge, for instance, obtained eight young eastern gorillas, all but one of which susbsequently died. The survivor, a young female mountain gorilla named Congo, was eventually studied by Robert Yerkes, who wrote a three-part monograph on the animal's intelligence which I shall refer to again in Chapter 5. Unfortunately, a high mortality rate has always accompanied the capture of gorillas. The usual practice of shooting whole groups of adults in order to take their infants has led to a great reduction in their numbers, particulary where the western lowland subspecies is concerned. This still continues today, for instance in Cameroun, where I visited a dealer who had five juvenile and infant gorillas awaiting sale to zoos. Gorillas have also long been hunted for food by various tribes such as the Fang of Rio Muni, who consider gorilla meat a delicacy. Animals are also killed when they raid plantations and any youngsters obtained may be sold to dealers. Many of them die soon after capture, sometimes as a result of the trauma of being separated from their mothers, sometimes from dietary causes, or from respiratory and other diseases. Even expert animal collectors have problems with the gorilla. Armand Denis, for example, lost thirty animals from disease in 1942.

The first live gorilla to reach Europe was exhibited in Wombwell's Menagerie during 1855, whilst the first gorilla seen in America arrived in 1898. London Zoo obtained gorillas in 1887, 1896 and in 1904 but all of these died quite soon after arrival. Similar failures were experienced by other zoos and even today, when air transport and expert veterinary care are available, animals may perish during the early stages of captivity. In 1953, Bernard Grzimek estimated that there were only fifty-three gorillas kept outside Africa and Cousins reported that thirty-eight of them were still living in 1967, a tribute to improving standards of care in zoos. It might be imagined that since the gorilla is a rare species, imports from the wild would have decreased. But in fact the reverse is true. Figure 1 shows the numbers of gorillas held in world zoos and other institutions between 1960 and 1978. During these years numbers in captivity rose steadily from 167 to 467.

Of captive gorillas held in 1978, 98% belonged to the western lowland subspecies, which has always been more heavily collected than the two eastern subspecies of gorilla. Zoos often classify all specimens from the eastern part of the gorilla's range as 'mountain' gorillas (*G. g. beringei*), whereas most of these in fact belong to *G. g. graueri*, the eastern lowland subspecies.

Since 1956 zoos have had increasing success in breeding gorillas, so that of the animals on display to the public in 1978, 102 were born in captivity. This is most encouraging and one may hope that in the future zoos will obtain all their gorillas from captive breeding programmes rather than from the wild.

Science has benefited greatly from studies of captive gorillas. Biochemical data have provided a foundation for new and controversial theories concerning the evolution of the great apes and man (see Chapter 4). Specimens which die in captivity have provided an opportunity to study the anatomy of the soft parts as well as the skeleton (Chapter 3). Many valuable observations have been made on aspects of the gorilla's behaviour and reproductive physiology as

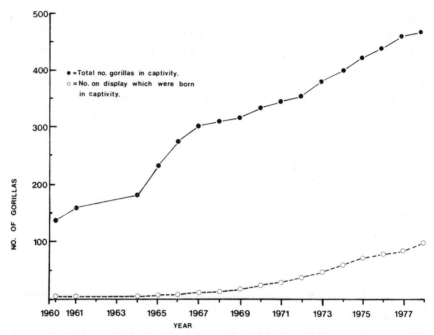

Figure 1. Numbers of gorillas in captivity in world zoos and other institutions between 1960 and 1978. Data are from International Zoo Yearbook Vols 3–19.

well as its growth and development (Chapters 3, 5, 6 and 7). It is depressing, therefore, that gorillas may become extinct in the not too distant future, not as a result of the traffic in captive gorillas, but because of the destruction of their natural habitat. Rain forests are being felled at an alarming rate in Africa, Asia and South America simply because the demand for timber and land is so great. Gorillas, like most primates, depend upon the forests for survival and unless areas are set aside and protected they will die out. This topic is discussed in the final chapter.

Chapter 2

Classification and Distribution

The primates

THE order Primates contains about 188 species and includes the prosimians, monkeys, apes and man. These animals are so diverse anatomically that no single characteristic will serve to distinguish them from other groups of mammals. Hence, the first comprehensive definition of the Primates, written by Mivart in 1873, was of necessity a broad one:

Unguiculate, claviculate placental mammals, with orbits encircled by bone; three kinds of teeth, at least at one time of life; brain always with a posterior lobe and calcarine fissure; the innermost digit of at least one pair of extremeties opposable; hallux with a flat nail, or none; a well developed caecum; penis pendulous; testes scrotal; always with two pectoral mammae.

Primate classification is not a straightforward matter because, as George Gaylord Simpson commented some years ago, it 'has become the diversion of so many students unfamiliar with the classification of other mammals that it is, frankly, a mess'. A plethora of recent publications has done little to alter the situation so, for the sake of simplicity, I shall follow Simpson's classification of 1945 which divides the primates into two suborders; the Prosimii and the Anthropoidea.

Modern prosimians are more similar, anatomically, to the ancestral primate stock of sixty or seventy million years ago than the anthropoid primates are. Hence they are sometimes, rather misleadingly, referred to as 'lower' primates whilst the anthropoids are called 'higher' primates. The lemurs, bushbabies and lorises of Africa, Madagascar and Asia are all prosimians and twenty-five of the thirty-seven prosimian species are nocturnal. All prosimians, except the tarsier, have a moist nasal area or rhinarium (see Plate 2) such as occurs in dogs or rodents, but which is lacking in monkeys and apes. Prosimians also possess a distinctive 'toilet claw' on the second digit of each hind foot. Until recently very little was known about the social organization of the nocturnal forms. Radio-tracking studies by Charles-Dominique, Bearder and Martin have shown that pottos and bushbabies have a basically 'non-gregarious' but complex social organization in which male home-ranges overlap the ranges of one or more females.

Two groups which have often been included in the Prosimii deserve special mention: the tree shrews and tarsiers. Tree shrews are squirrel-like creatures found in various parts of Southeast Asia. Many authorities consider that they should not be classified as primates. The tarsiers are tiny nocturnal primates which feed on lizards and other small animals. They occur in Borneo, Celebes and the Philippines and rarely survive long in captivity. We know very little about their behaviour, but recent fieldwork by John Mackinnon indicates that they live in monogamous family groups. Anatomically, the tarsiers have been exhaustively studied, for they are in many ways structurally intermediate between prosimians and anthropoids. How the tarsiers should be classified is still debated and some place them in a suborder of their own.

The anthropoid primates are divided into three superfamilies, the first of these being the Ceboidea or New World primates. New World primates are found in Central and South America and are also called platyrrhines because their nostrils are widely spaced and point sideways. By contrast, the Old World monkeys, apes and man are often called catarrhines, because their nostrils are close together and point downwards (see Plate 2).

The platyrrhines include the tiny marmosets and tamarins (family Callitrichidae) and a large number of more 'monkeylike' forms such as woolly, howler, spider and capuchin monkeys (family Cebidae). With the exception of the owl monkey (*Aotus*) these are diurnal animals and all of them are arboreal. Many of the Cebidae are superficially very similar to the Old World monkeys of Africa and Asia. This may be partly attributable to parallel evolution, which has occurred in response to similar selective forces. South American monkeys have adapted to life in the trees, feeding on vegetation or insects and living in social groups just as Old World monkeys do. South American primates have three premolar teeth in each half of the jaw as compared to two in catarrhines and some, such as spider or woolly monkeys, have prehensile tails. These are only two of many structural features which distinguish New World primates from their relatives of Africa and Asia.

The catarrhine primates are divided into two superfamilies: the

Cercopithecoidea, or Old World monkeys, and the Hominoidea, which includes the apes and man. The monkeys of Africa and Asia represent a wide range of arboreal forms such as *Colobus* and *Cercopithecus* and also more terrestrial types such as geladas, baboons and macaques. The hominoids include the lesser apes (gibbon and siamang), the great apes (gorilla, chimpanzee and orang-utan) and man.

Evolutionary 'trends' among primates

Sir Wilfred Le Gros Clark developed the concept that certain 'trends' have occurred during primate evolution and that the existing members of the order may be arranged in a graded series which demonstrates the increasing expression of these trends. The series begins with prosimians and proceeds to the monkeys, apes and man (Fig. 2).

Primates have shown a tendency to retain certain features such as clavicles, pentadactyl limbs and a simple pattern of ridges or 'cusps' on the molar teeth. Features which are advantageous for living in trees (and perhaps for preying upon insects) have been selected for; hence the development of eyes which point forward and which make stereoscopic vision possible. Hands and feet which grasp are also useful in the trees and so primates have tended to develop an opposable

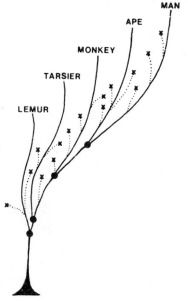

Figure 2. Diagram showing Le Gros Clark's scheme for arranging primates in a graded series, from lemurs to man. The series suggests a general 'trend' in evolutionary development. Xs represent postulated ancestral forms. Redrawn and adapted from Le Gros Clark (1959).

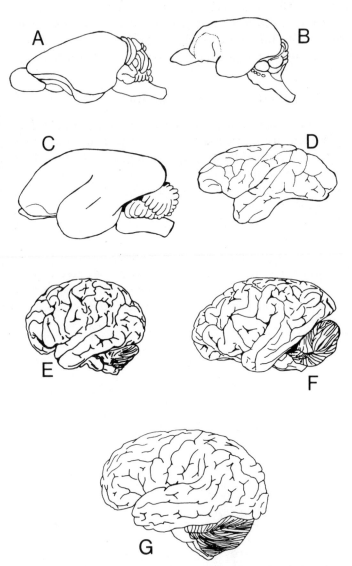

Figure 3. Lateral view of the brain of a tree shrew and of various primates (not drawn to scale).　**A** Tree Shrew; **B** Tarsier; **C** Marmoset; **D** Macaque; **E** Orang-utan; **F** Gorilla; **G** Man. Gorilla and Orang-utan; author's specimens and drawings. Remaining figures; redrawn from Napier and Napier (1967) and Le Gros Clark (1959).

thumb and big toe, and have replaced claws with flattened nails and sensitive tactile pads on the ends of the digits.

Methods of nourishing the developing foetus have also changed during primate evolution and the placental circulation of tarsiers and anthropoid primates differs from that of prosimians. Marked changes have also occurred in the brain. The olfactory bulbs, which serve the sense of smell, have shown a tendency to reduce in size relative to the cerebral hemispheres, which have gradually enlarged and become folded into a complex pattern of gyri and sulci (Fig. 3).

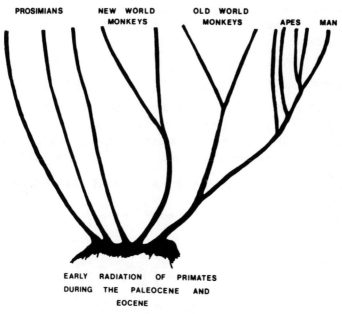

Figure 4. A simplified scheme of the evolutionary affinities of extant primates. The lengths of the branches are approximate and do not represent a time scale.

Such generalizations may be misleading; the size of some portion of the brain is not a sure guide to its functional importance. The fine structure and physiology of the neural tissue is more relevant in this respect. However, it seems probable that olfaction plays a more important role in the social life of prosimians and New World monkeys than in Old World monkeys or apes. The accessory olfactory system (Jacobson's organ) is rudimentary or absent in catarrhine primates, except during foetal life, whereas prosimians and platyrrhines have a well-developed Jacobson's organ and accessory olfactory bulbs. It is likely, but not proven, that these are concerned with perception of olfactory cues which elicit sexual responses.

The series of brains in Fig. 3 shows a gradual increase in the size of the cerebrum and decrease in the olfactory region as exemplified by living primates. The gorilla stands near the apex of the series and the complexity of its cerebral development is surpassed only by man.

Although the concept of evolutionary trends is useful, enabling us to look at the gorilla in relation to other primates, it can be misleading. The kinds of relationship shown in Figs 2 and 3 do not illustrate an evolutionary descent: the more 'complex' species did not evolve from the less complex ones. The different evolutionary lines arose in the tertiary period and have subsequently developed quite separately into today's prosimians, platyrrhines and catarrhines, in a fashion shown in simplified form in Fig. 4. Each separate branch has taken a separate evolutionary route and developed its own gradations and degrees of complexity.

The apes

Traditionally, the apes have been divided into the 'lesser' apes of the family Hylobatidae (gibbon and siamang) and the 'great ' apes or Pongidae (gorilla, chimpanzee and orang-utan) (Table 1). Gibbons (see Plate 3) are widely distributed in Southeast Asia and the sexes

Table 1. Classification of the apes and man (superfamily Hominoidea

Family	Genus	Species and subspecies	Common name
Hylobatidae	*Hylobates*	* *H. lar.*	Gibbon
	Symphalangus	S. *syndactylus*	Siamang
		S. s. *syndactylus*	
		S. s. *continentis*	
Pongidae	*Pongo*	*P. pygmaeus*	Orang-utan
		P. p. pygmaeus	Bornean orang-utan
		P. p. abelii	Sumatran orang-utan
	Pan	*P. troglodytes*	Chimpanzee
		P. t. troglodytes	
		P. t. verus	
		P. t. schweinfurthii	
		Pan paniscus	Pigmy chimpanzee
	Gorilla	*G. gorilla*	Gorilla
		G. g. gorilla	Western lowland gorilla
		G. g. beringei	Mountain gorilla
		G. g. graueri	Eastern lowland gorilla
Hominidae	*Homo*	*H. sapiens*	Man

* The type species is *H. Lar*. Many other species and subspecies of gibbon have been described by various taxonomists.

are roughly equal in size (weighing four to seven kilogrammes) in contrast to the pongids, where males outweigh females. Gibbons have strikingly elongated arms and hands and are the most agile exponents of brachiation, hanging by their hands and swinging arm over arm from branch to branch. The siamang is a larger (nine to twelve kilogrammes), longer-armed and shaggier version of the gibbon, and both sexes have a throat sac which is inflated during some of their vocal displays. Siamangs are found only in Malaya and Sumatra.

The great apes occur in Southeast Asia (orang-utan) and in Africa (gorilla and chimpanzee). Orangs (see Plate 4) were once more widely distributed than they are today; the fossilized teeth of orang-utans have been found on the mainland of China. They now only survive on Borneo (*Pongo p. pygmaeus*), and in Northern Sumatra, where a larger subspecies (*P. p. abelii*) with lighter, reddish-brown hair is found. Orangs are large arboreal apes, males weighing about seventy

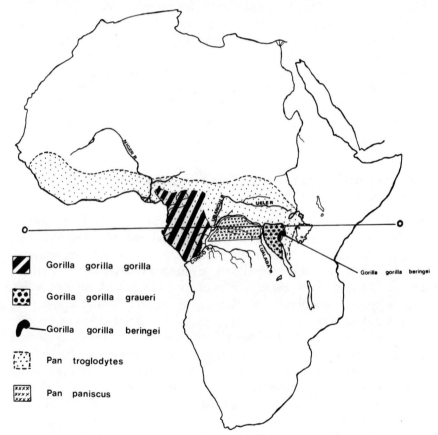

Figure 5. Recent limits of distribution of the gorilla and chimpanzee. Based upon Schaller (1963); Schultz (1969); Groves (1970) and Jantschke (1975).

17

kilogrammes and females thirty-seven kilogrammes. Although quite capable of brachiating, they usually climb more slowly, gripping the branches with enormous elongated hands and feet. Old males have big cheek flanges consisting of fat and fibrous tissue.

The chimpanzee (*Pan troglodytes*) (see Plate 5), is divided into three subspecies and is widely distributed across forested equatorial Africa (Fig. 5). A second species of chimpanzee, *Pan paniscus* the pigmy chimp, is restricted to a small area in the Congo Basin. Its existence is not in doubt, for there are specimens in zoos and museums. Fieldwork is in progress to study its ecology and behaviour. Pigmy gorillas have not been proved to exist, although they have been reported as being four to five feet tall with greyish brown hair. The first such gorilla (*G. mayema*) was described by Alix and Bouvier in 1877 from the Congo (Brazzaville). This, and later descriptions by Elliot and Frechkop, are now considered to be mistaken identifications of ordinary female gorillas or male chimpanzees.

The African apes, and gorillas in particular, are much more terrestrial in habit than are the gibbons, siamang or orang-utan. Although the ancestors of African apes were probably more arboreal in behaviour, the gorilla and chimpanzee have become modified for a special kind of terrestrial locomotion called 'knuckle-walking'. They rest their weight on the backs of their flexed fingers while walking on all fours, rather than putting their hands down flat on the ground as baboons or macaques do (see Plate 6).

Possible modifications to the classification of the Hominoidea

Table 1 summarizes a scheme for the classification of the Hominoidea by dividing it into three families: the Hylobatidae (lesser apes), Pongidae (great apes) and Hominidae (man). Primate taxonomy is not static, however, and a bewildering number of alternative schemes are used in textbooks. Firstly, opinions vary as to how many species and subspecies of gibbons and siamangs there are. Elliot, for instance, considered that there are ten species in this family, whilst Pocock listed only three. Recently Colin Groves has grouped the gibbons and siamang into a single genus, Hylobates, containing six species and eighteen subspecies. He includes this genus in the family Pongidae.

The pigmy chimpanzee has been given specific rank in Table 1, but some authorities do not think that this is warranted, and classify it as a subspecies of *Pan troglodytes*. A more important problem concerns whether chimpanzees and gorillas should be placed in the same genus. Recently this view has become popular and the gorilla is often referred to as *Pan gorilla*. This is not a modern concept, for, when the gorilla was discovered in 1847, Savage and Wyman placed it with the chimpanzee, in what was then the genus *Troglodytes*. Since then the gorilla has been given a number of generic titles, including *Pithecus*, *Anthropithecus* and *Satyrus*. However, as will become clear during the course of this book, there are many differences in anatomy, ecology and behaviour between chimpanzees and gorillas. In recognition of this fact, and also to avoid the confusion which results from constantly changing the Latin names of animals, it seems best to retain the two apes in separate genera.

It has been suggested that gorillas and chimpanzees may interbreed in the wild. Illustrations of possible hybrids called 'kooloo-kambas' are included in some papers by Duckworth and in Yerkes and Yerkes treatise, *The Great Apes*. It is claimed that kooloo-kambas have been observed in normal groups of chimpanzees, which suggests a cross between a male gorilla, perhaps a lone silverback, and a female chimpanzee. Such a cross is most unlikely, however, since male gorillas may weigh over 150 kilogrammes; this is more than four times the weight of a female chimpanzee. Moreover, the genital morphology and sexual behaviour of the gorilla differ so greatly from those of the chimpanzee that it seems very unlikely that these apes would hybridize in the wild. A cross between the two might be feasible in captivity, if artificial insemination were used, but I have not heard of such an experiment.

Finally, there is the question of whether the African apes should be placed with man in the family Hominidae, whilst retaining the orang-utan in the Pongidae. This suggestion is not as odd as it seems at first sight, for the gorilla and chimpanzee are in many respects more similar to man than the orang-utan is. *Homo sapiens* has always been placed in his own family and many anthropologists argue that, despite his kinship with gorillas or chimpanzees, man is sufficiently unique to merit such a classification. He is the only habitually bipedal primate and shows many modifications of the pelvis, vertebral column, legs and feet in association with an upright gait. Most important, also, are man's mental abilities, for he has acquired a capacity for language and abstract thought which has propelled him into a cultural orbit far beyond that of apes. However, language capacity and tool-making are not uniquely human attributes: chimpanzees possess both these abilities, albeit at a very basic level, as we shall see in Chapter 5.

History of the classification of the gorilla

There is one species and three subspecies of gorilla; the western lowland *G. g. gorilla*, the eastern lowland *G. g. graueri* and the rare mountain subspecies *G. g. beringei*. Mountain gorillas differ from the western lowland animals in various ways (Table 2). Mountain gorillas have longer and blacker hair. They also have a narrower skull, longer palate, shorter arms and broader hands than the western subspecies. The foot of the mountain gorilla differs slightly from that of the western gorilla in that the big toe is more closely aligned with the other toes (Fig. 6). This feature correlates with the observation that

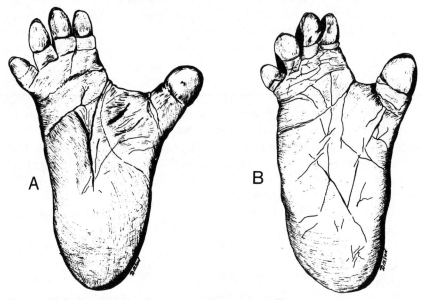

Figure 6. Right feet of adult male western and mountain gorillas.
A *Gorilla g. gorilla* (author's drawing); **B** *Gorilla g. beringei* (redrawn from Schultz (1969).

mountain gorillas are probably the most terrestrial of the three subspecies. The scalp of the adult male *G. g. beringei* often has a mitre-like appearance due to an underlying pad of fibrous tissue which is not nearly· so prominent in the western form (Fig. 7). Whereas the western lowland animal has a 'lip' or projection above the nasal septum, this is lacking in *G. g. beringei*. The eastern lowland gorilla typically has a longer narrower face, but in general it is intermediate in structure between the other two subspecies. The reasons for this will become clear later, when the past and present distribution of gorillas is considered.

Figure 7. Portraits of an adult male western lowland gorilla (left) and a mountain gorilla (right). Author's drawings from photographs.

Gorillas were originally considered as a single species, for instance by Forbes in his *Handbook to the Primates* published in 1897. Very few specimens had been collected then and little was known about the animal. The gorilla shot by Von Beringe in 1902 in the Virungas was classified by a German taxonomist, Matschie, as a new species, *G. beringei*. In the years which followed, as more gorillas were killed and brought back to European museums, many more 'species' were described by Matschie, such as *Gorilla diehli*, *G. matschei*, *G. jacobi*, and *G. graueri*. Today, the minor differences between these animals are not considered sufficient to warrant dividing them into species. Gorillas show considerable individual variations, age changes, sex differences in hair colour, skull morphology and other features. This was not appreciated in Matschie's day. For instance, young male gorillas lack the saddle of white hair on the back and the large crests on the roof of the skull which develop as they reach adulthood and, again, these are much smaller or absent among females. Once such variations are understood, some of the taxonomic characters selected in the past begin to appear trivial. Elliot, in his 1913 *Review of the Primates*, lists features such as 'legs below knees black ... with a chestnut patch on the head' and 'without a beard' in his key for classifying two gorilla species and four subspecies. In fairness, he expressed 'sincere doubt regarding the status given examples' but considered that 'the material in the museums of the world is not sufficient to prove that any one of them is not entitled to the rank assigned to it'. Similar problems frustrated earlier attempts to classify the orang-utan, which also exhibits marked sex differences in size and much individual variability.

In 1929, Harold Coolidge published a monograph entitled *A Revision of the genus Gorilla*. His studies involved measurements of skulls in many museums and private collections. The mathematical treatment he gave his data has been criticized and revised subsequently but his conclusion, that there is only a single species of gorilla, remains unchallenged. He recognized two subspecies; the western lowland *G. g. gorilla* and the eastern or mountain *G. g. beringei*. Coolidge, and later Adolph Schultz, listed many differences between eastern and western gorillas, the most important of which are shown in Table 2. Schultz examined eastern gorillas mainly from

Table 2. Some differences between the western lowland gorilla (*Gorilla g. gorilla*) and the mountain gorilla (*G. g. beringei*)

Compared to the western subspecies, the mountain gorilla has:

1 Longer, thicker, black hair
2 Narrower skull
3 Longer palate
4 Fleshy 'callosity' on top of head (in adult males)
5 No 'lip' above the nasal septum
6 Shorter limbs (particularly the arms)
7 A more curved vertebral border to the scapula
8 Shorter, broader hands
9 Great toe branching from the sole more distally
10 Webbed toes (usually)
11 Greater trunk length
12 Narrower hips

Information from: Coolidge (1929), Schultz (1934), Groves (1970).

the Virunga volcanoes and considered that these 'short-armed gorillas' should be classified as a second species. This population is still classified as *G. g. beringei*, however, whereas the majority of eastern gorillas have recently been placed, by Groves, in a new subspecies, *G. g. graueri*.

After Emlen and Schaller conducted their field studies on eastern gorillas in 1959, and reported that most of them lived in low-altitude rain forest rather than in mountainous areas, a further study of variability in gorillas was made by Colin Groves of Cambridge University. He made the important advance of relating anatomical variation between gorilla populations to differences in their ecology. Western lowland gorillas do not only live in low-lying areas; they have been found at 1,500 metres in the Cross River area of Nigeria and at 600 metres in parts of Cameroun. Groves divided western lowland gorillas into four groups; coastal, plateau, Sangha and Nigerian (Fig. 8). Two-thirds of eastern gorillas are found in lowland forests at between 490 and 800 metres. The remainder do live in highland areas and one small population occurs up to 3,900 metres in the Virunga volcanoes. Groves also divided eastern gorillas into

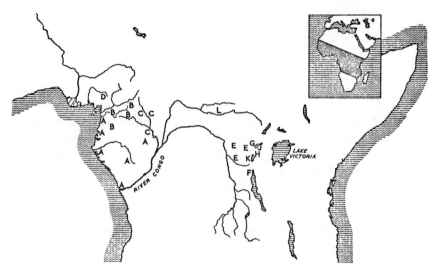

Figure 8. Map to show the divisions used by Groves in his study of subspecific variability in the gorilla.

    A *Gorilla g. gorilla* (coast); **B** *G. g. gorilla* (plateau); **C** *G. g. gorilla* (Sangha); **D** *G. g. gorilla* (Nigeria); **E** *G. g. graueri* (Utu); **F** *G. g. graueri* (Mwenga-Fizi); **G** *G. g. graueri* (Tsiberimu); **H** *G. g. beringei* (Virunga); **J** Kayonza forest; **K** Mt Kahuzi.

After Groves (1967).

four groups; Utu, Mwengas-Fizi, Tsiaberimu and Virunga. He measured groups of skulls from adult males and females taken from all of these areas. Originally forty-five measurements were made on each skull but since many of these were positively correlated, the number was reduced to sixteen; ten measurements of the cranium and six of the lower jaw. The ecological diversity of the various groups was matched by differences in their skulls. Among western gorillas, Sangha animals were found to be intermediate between those of the coast and plateau regions, whereas Nigerian gorillas formed a more distinct group. The differences were small, however, and all these types truly belong to one subspecies, *G. g. gorilla*.

Among eastern gorillas, those of the Virungas stood out as a distinct group whilst the remainder were found to resemble one another, but with varying degrees of similarity to either western lowland gorillas or to the Virunga volcanoes population. This intermediate group is most easily distinguishable from the mountain form by its shorter hair and long narrow face. Groves first called this new subspecies *Gorilla g. manyema* but, following a criticism regarding taxonomic priority by Corbet, it was renamed *G. g. graueri*, after Robert Grauer, who shot specimens in the mountains west of Lake Tanganyika during the early part of this century.

True mountain gorillas (*gorilla g. beringei*) are to be found in two areas, the Virungas and Kahuzi. In fact, there are not enough skulls available to be certain how the gorillas of Kahuzi should be classified. In 1975, Michael Casimir of the Max Planck Institute in Berlin reported on some male crania and two mandibles from Kahuzi gorillas. The crania were indistinguishable from *G. g. beringei* specimens of the Virunga volcanoes, but certain measurements of the mandibles were more similar to *G. g. graueri* and Casimir suggests that the gorillas of Kahuzi should be placed in the eastern lowland subspecies. Groves, however, had placed the gorillas of Kahuzi in the mountain subspecies on the basis of studies of post-cranial skeletal material as well as skulls, and his scheme seems preferable.

The past and present distribution range of gorillas

Figures 5, 9 and 10 show the recent distribution range of the gorilla. Fig. 9 is based on information from museum collections and field studies. Each dot on the map records where gorillas have been observed or shot. It is more accurate than Fig. 5, on which entire areas are shaded over as if gorillas occurred throughout them. This is not so, for the species has a 'spotted' distribution and is found in some tracts of forest but not others. This is clear if we look at the map made by Schaller and Emlen of the distribution of eastern gorilla populations (Fig. 10). Eastern gorillas are found within a large area which stretches for about 220 miles from east to west and 300 miles from north to south. Within this area, however, the animals only occur in about sixty rather isolated forests. Emlen and Schaller have estimated that there may be 5,000–15,000 gorillas in these forests. *Gorilla g. graueri* is a creature mainly of lowland rain forested areas such as Utu, but is also found in mountainous areas like Tsiaberimu. *G. g. beringei* occurs only on six of the eight Virunga volcanoes and also at Kahuzi.

There is not enough information on the western lowland gorilla to discuss its distribution range in detail. In the 1950s, Blancou suggested that there might be between 10,000 and 20,000 western gorillas. Approximate estimates are the best that can be offered because

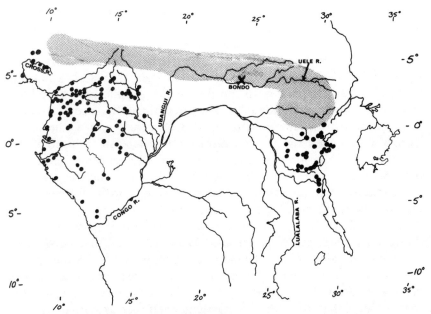

Figure 9. Recent distribution range of the gorilla and the extent of its postulated past distribution range.
    Solid circles: localities of collection of museum specimens or observation of living gorillas.
    Stippled band: hypothetical former distribution of the gorilla when the montane 'forest
            bridge' was in existence.
Redrawn and modified from Groves (1971).

very little fieldwork has been done and it is very difficult to census animals which live in dense forest. Some of the points on Fig. 9 mark areas where the subspecies is now extinct. In Cameroun, Merfield and Raven shot gorillas in the forests around Yaoundé in the 1920s, but there are certainly none there now. Although skulls of gorillas from the Cross River region of Nigeria may be found in various museums, this population was reported to be on the decline twenty years ago and it is now extinct.

The eastern limits of the range of *G. g. gorilla* occur in the Central African Republic and a gap of about 1,000 kilometres separates western lowland gorillas from the area occupied by the two eastern subspecies (Fig. 9). Why do no gorillas live in the immense intervening forests of the Congo basin? The three subspecies are quite similar, which would indicate that they diverged fairly recently. Not too long ago, western and eastern gorillas must have had a continuous distribution range.

Emlen and Schaller have pointed out two facets of gorilla behaviour which must have been important in limiting their distribution range.

25

Firstly, they shun open areas and stick mainly to the cover of rain forest. Secondly, they cannot swim and large rivers constitute an impassable barrier to them. Casimir has observed gorillas feeding in marshy areas at Kahuzi, but this is exceptional. Gorillas usually avoid water and if they have to cross a stream then they seek a natural bridge, such as a fallen log. In captivity, when gorillas have been kept in an enclosure surrounded by a moat, animals have occasionally fallen in and drowned. As can be seen in Figs 5 and 9, the River Congo (now called the River Zaire) marks the eastern limit of the distribution of *Gorilla g. gorilla.* This river is the seventh longest in the world, and is fed from the north by tributaries such as the Uele and Ubangui. A water barrier therefore bars the gorillas from spreading into the rain forests of the Congo basin.

If we look at the distribution range of eastern gorillas (Fig. 10), they too are restricted by physical barriers. In the south, at Fizi, open savannah and agricultural development begins and limits the spread of *G. g. graueri.* In the east, there is the Rift escarpment and chain of great lakes. Large open areas of grassland occur here and it is possible that these, rather than the escarpment itself, act as barriers to the gorillas. The gorillas of Kahuzi and the Virunga volcanoes perhaps reached these areas by way of forests which have since been destroyed by agriculture. To the west of present gorilla distribution is the Lualaba River, but gorillas cease to populate the forests well to the east of the river, and Emlen and Schaller do not consider that it acts as a barrier. Instead, they believe that gorillas have only quite recently spread southward into this part of Africa, around the head-waters of rivers such as the Oso and Lugulu, and are at present still spreading westward towards the Lualaba. The northern boundaries of *G. g. graueri* shown in Fig. 10 are only approximate, for not enough fieldwork has been done to define them.

To understand how western and eastern gorillas were linked together in the past, it is necessary to realize that the climate of Africa and the distribution of its forests has changed during recent geological time. Moreau considered that, in the late Pleistocene period, when the climate was cooler, forest occurred along the northern edge of the Congo basin in areas which are not suitable for gorillas today. The Congo basin itself was occupied by a massive lake, called Lake Busira. Emlen and Schaller suggest that gorillas spread from the west by way of the 'forest bridge' round the northern edge of the Congo basin, and thence down into their present range in Uganda, Rwanda and eastern Zaire (Fig. 9). When the climate changed, the area occupied by the 'forest bridge' became hotter and drier, the forest retreated and contact between western and eastern gorillas was broken.

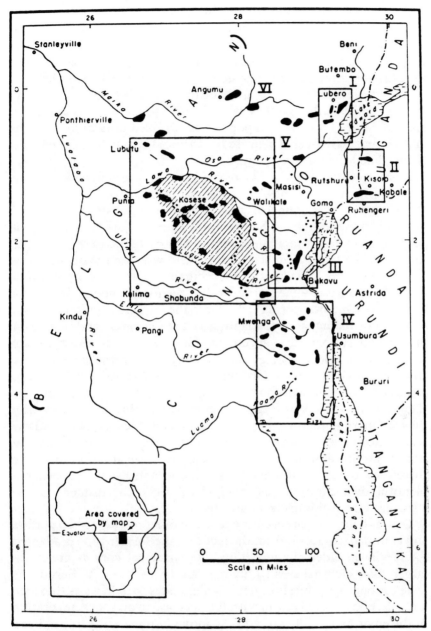

Figure 10. Map showing the recent distribution of eastern gorillas.

I Mt Tsiaberimu region, II Virunga volcanoes, III Mt Kahuzi region, IV Mwenga-Fizi region, V Utu region, VI Angumu region.

The populations of the Virungas and Kahuzi belong to *Gorilla g. beringei*. The remaining populations are classified as *G. g. graueri*. From Schaller (1963), after Emlen and Schaller (1960).

Further evidence in support of this idea is available in the form of four gorilla skulls which have been collected from Bondo in the Uele Valley, an area where gorillas no longer occur. These skulls, two from specimens shot by a soldier and two obtained from native villagers, were first described by Schouteden in 1927 as a new subspecies, but Groves considers that they are indistinguishable from the western lowland gorilla although they were found 650 kilometres outside its present range. Thus, a small population survived until quite recently in what may have been a plentifully inhabited area before the climate changed and the forests diminished.

More recently, Groves has reconsidered the problems of where gorillas originated and how they reached the areas they occupy now. He believes that the animal originated in the east and spread westward rather than the reverse, as Schaller postulated. Groves considers that gorillas originated in montane forest, such as is found in the Virungas today. In the late Pleistocene the climate was cooler and montane forest was much more widespread, occurring round the northern part of the Congo basin – the 'forest bridge' mentioned above. When the climate changed, the gorilla spread into the lowland rain forest which is its major habitat today, adapting to the changed conditions as it did so. To support this argument, Groves reminds us of certain features of the gorilla ecology and anatomy. Firstly, the largest group sizes and highest population densities recorded are for *Gorilla g. beringei* on Mount Mikeno in the Virungas. The montane forests there contain abundant forage for gorillas and this is probably why they are more numerous. Lowland rain forest is less suitable and the scant information available indicates that lowland gorillas live in smaller groups and have a lower density of population. Gorillas are therefore more successful in montane forests and also tend to be more plentiful than chimpanzees in highland areas, for instance in the Virungas where chimpanzees are absent.

Groves believes that certain anatomical differences between gorillas and chimpanzees result from the fact that the former type is adapted for life in a colder climate. Thus, for instance, gorillas are much larger than chimpanzees. This may be an example of Bergman's 'rule', a zoogeographical principle which states that, in warm-blooded animals, body size increases with the decrease in average temperature of the enviroment. This trend is adaptive in preventing loss of body heat, for the larger an animal becomes, the greater is its mass in relation to body surface and the better its ability to retain heat in cold conditions. The expanded nostrils of the gorilla and its broad deep chest may also be adaptations for breathing colder air. Gorillas also have shorter ears, penis and limbs than do chimpanzees, which,

Groves believes, may be examples of Allen's rule. This states that, in warm-blooded animals, the extremities decrease in size in colder climates and that this, again, is an adaptation connected with heat regulation. A classic example of Allen's rule concerns ear length in foxes; the artic fox has very short ears, whereas the ears of European foxes are of medium length. The desert fox, living in the hottest climate of the three, has by far the longest ears. The anatomical adaptations to cold climate shown by gorillas are most pronounced in the mountain subspecies of the Virungas which, in addition, has longer hair than the lowland types.

It is possible, then, that gorillas spread across to western central Africa in the late Pleistocene via a montane forest bridge. In Groves's opinion, the eastern lowland gorilla is more distinct from the mountain form than it is from the western lowland subspecies. He suggests that this indicates that the two eastern subspecies diverged farther back in time and that gorillas must, therefore, have originated in the eastern part of their range. However, evolution does not always occur at the same rate in different populations. The mountain gorillas of the Virungas might have spread to this area more recently and might have rapidly evolved their distinctive characteristics in respose to the extreme environmental pressures there. It has also been pointed out by Donald Cousins that, if gorillas originated in cool montane forests, then it is surprising that they are so sensitive to respiratory diseases such as pneumonia. Whatever the truth of the matter, however, Groves's hypothesis is certainly an elegant explanation of the facts.

Chapter 3

Structure and Function

MORE has been written about the structure of the gorilla than about any other aspect of its biology. There are several reasons for this. Traditionally, studies of morphology and anatomy have occupied a key position in biology. Statements about how an animal functioned, how it should be classified and its evolutionary affinities were based largely upon studies of its structure after death. Interest in the gorilla, because of its close relationship to man and the scarcity of living specimens for study also contributed to a bias in favour of anatomical researches.

To review all the literature on the structure of the gorilla is a task for an expert and would require a volume to itself. I can only offer a basic account which provides a useful background to other chapters. For detailed information, the reader may refer to the bibliography and in particular to work by Duckworth, Raven and Sonntag or to the numerous valuable articles by the late Professor Adolph Schultz.

The gorilla's hair and skin

The gorilla's hair is most frequently black, particularly in the two eastern subspecies. Western gorillas, however, show much variability and the hair may contain various shades of brown or even reddish tints, particularly on the scalp. Hair follicles occur in groups of three to five and are densest on the top of the head, on the upper arms and back. Hair is very sparse on the face, chest and in the armpits and is thinner on the belly than on the back. As I mentioned in the last chapter, mountain gorillas have longer hair than the two lowland subspecies, presumably as an adaptation to their colder environment.

As a male gorilla reaches adulthood the short hair on his back begins to turn grey. Gradually a prominent greyish saddle develops which has given rise to the name 'silverback'. This feature is particularly striking in male mountain gorillas in which the short grey hair on the saddle contrasts with the long shaggy black coat on the limbs, shoulders and head. As silverbacks get older, particularly members of the western subspecies, the grey areas spread to include the flanks and buttocks (see Plate 6). Tales of 'white gorillas' in western central Africa may originate from sightings of such old, grey males. Only one white gorilla has actually been captured and is kept at Barcelona Zoo. This male is blond and not albino as is often stated, for he has blue eyes and not pink ones as albinos do.

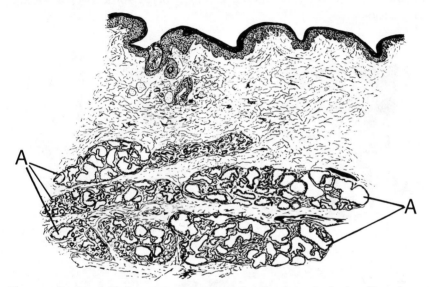

Figure 11. Section of skin from the armpit of an adult male western lowland gorilla to show the 'axillary organ'.

A Groups of large apocrine glands. Author's specimen and drawing.

A few grey hairs may be found scattered throughout the coat of some female and infant western gorillas. Infants may also possess a tuft of white hairs on the rump, a feature which is also found in young chimpanzees. This white tuft usually turns black as the infant gets older but I have seen some female western gorillas in which these white rump hairs have persisted into adulthood.

The gorilla's skin is black. In some baby gorillas, pigment may be missing from portions of the fingers and toes or from the palms and soles so that they appear pink. The unpigmented areas have usually darkened, however, by the time the youngsters reach the juvenile stage. In adults the skin is thickest on the back and thinnest on the naked chest and in the armpits. Two kinds of sweat glands are found in the skin, the apocrine and eccrine glands. Straus has pointed out that in adult males the armpits contain many large apocrine glands arranged in four to seven layers. This so-called 'axillary organ' (Fig. 11) is responsible for the rank odour of siverbacks, often remarked upon by gorilla hunters and fieldworkers. Female gorillas do not possesss such a well-developed axillary complex or such a pungent smell as males. It is interesting to note that an axillary organ similar to that of the gorilla also occurs in the chimpanzee and in man but has not been reported either in the orang-utan or gibbon. The function of the axillary organ remains obscure though it has been suggested that the female's odour plays some role in sexual interactions.

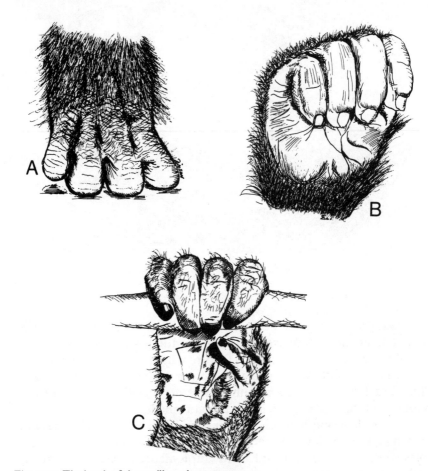

Figure 12. The hands of the gorilla and orang-utan.
    A Hand of a young gorilla in a knuckle-walking posture. In this example the fifth finger is not in contact with the ground.
    B Adult male gorilla's hand showing the 'knuckle pads'.
    C Hand of an adult female orang-utan in a brachiating posture.
Author's drawings: not to scale.

Eccrine sweat glands are found in many areas of the skin, but the highest concentrations are found in the palms of the hands and soles of the feet. The epidermis in these regions is very thick and consists of complex patterns of ridges and furrows called dermatoglyphics. In the gorilla, just as in man, these dermatoglyphics serve to improve grip and sensitivity and the eccrine glands have an important lubricating function. The gorilla 'knuckle-walks', by resting its weight on the backs of the middle joints of the flexed fingers (Fig. 12). The skin on the backs of these joints is greatly thickened and Ellis and Montagna have shown that it also contains large numbers of eccrine glands.

Body weight

It is obviously no easy task to weigh a wild gorilla, and for this reason few gorilla hunters ever recorded accurate weights for their trophies. Some sportsmen were tempted to make up for the lack of facts with a few imaginative guesses. One Frenchman, for instance, claimed that a male gorilla he shot in 1905 weighed 350 kilogrammes and stood 2.3 metres high! The most reliable measurements of wild-shot, adult male gorillas have been collected together by Groves, and some of them are shown in Table 3. The average weight of adult males of all three gorilla subspecies is 152.77 kilogrammes. There seem to be some subspecific differences; eastern gorillas are heavier than western lowland ones and the heaviest animals of all belong to *G. g. graueri*. The information on *G. g. graueri* only refers to two males, however, so it may not be representative.

Adult female gorillas weigh only about half as much as males. This applies to gorillas in zoos as well as to wild specimens. Information on captive animals must be treated carefully because gorillas may become obese in captivity. Take, for instance, the case of 'Phil', a male western gorilla who was kept at St Louis Zoo from 1941 until 1958. After six weeks on a diet 'Phil' still turned the scales at 352 kilogrammes, more than twice what a healthy male should weigh! Figure 13 shows body weights of captive male and female gorillas from birth to fourteen years of age. The graph is useful as a general guide, but some of the points on it may be inaccurate since they are averages for only a few animals. I cannot vouch for the accuracy of weights published by various authors, but values which seemed exceptionally high, indicating obesity, or unusually low, perhaps because the gorilla was sick or its age wrongly estimated, were excluded.

Both male and female gorillas weigh only about 2.1 kilogrammes at birth. For the first six years of life the sexes increase in weight at a similar rate, but between six and seven years of age males begin to

33

Table 3. Some body weights and measurements of adult male gorillas

Subspecies	Weight (kg)	Height (cm)	Girth (cm)	Arm-span (cm)	Arm length (cm)	Leg length (cm)
G. g. gorilla	139.4 (6)	168.5 (25)	143.0 (19)	233.7 (17)	111.6 (8)	76.8 (8)
G. g. graueri	163.4 (2)	175.0 (4)	152.3 (3)	259.5 (2)	114.0 (2)	79.0 (2)
G. g. beringei	155.5 (13)	172.5 (6)	146.7 (13)	227.5 (10)	106.0 (4)	76.3 (3)

Number of specimens given in parentheses.
Data from Napier and Napier (1967), after Groves.

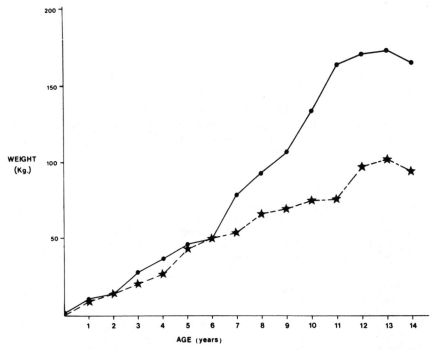

Figure 13. Body weights of captive male gorillas (dots) and females (stars) from birth until fourteen years of age.

Data are averages. Principal sources: *Annual Reports of Jersey Wildlife Preservation Trust* and *International Zoo Yearbooks*.

put on weight much faster than females. In fact, Gijzen and Tijskens have shown that both sexes exhibit a spurt of growth during puberty, which occurs at about six and a half years of age in females and perhaps a little later in males. Similar adolescent growth spurts have been demonstrated only in higher primates such as the rhesus monkey, chimpanzee and man. In all cases, males show a more pronounced growth spurt than females do.

Adult male gorillas aged between ten and fourteen years weigh on average 162 kilogrammes whilst females in the same group average 89 kilogrammes. Why are male gorillas so much bigger than females? Among the other hominoids only the orang-utan shows this trait to the same extent: females average 48% of the males' weight. In the chimpanzee, however, females are 90% as heavy as males and in the gibbon and man the sexes do not show extreme weight differences. Orang-utan males are fairly solitary creatures and the big old males are territorial, so it is possible that their massive size and striking secondary sexual adornments (see Plate 4) have been selected for on the basis of intermale competition for territory and mates. Gorillas,

35

however, live in groups and occupy home ranges. They do not defend territories and though conflicts between adult males certainly do occur, they seem to be rare. Since there is often only one silverback in a group of gorillas, and since, unlike the orang, gorillas are terrestrial animals, they may be more vulnerable to attack by predators. It is possible, therefore, that the male gorilla's massive bulk is also a reflection of his role as defender of the social group. Silverback males may also spend long periods away from groups and presumably large lone males are less likely to be attacked than a smaller ape would be.

Table 4. A comparison of various body proportions in macaques, apes and man

Proportion	Macaque	Gibbon	Orang-utan	Chimpanzee	Gorilla	Man
$\dfrac{\text{arm length}}{\text{leg length}} \times 100$	112	165	172	136	138	80
$\dfrac{\text{arm length}}{\text{trunk length}} \times 100$	113	243	200	172	172	148
$\dfrac{\text{chest breadth}}{\text{chest depth}} \times 100$	88	117	126	127	138	128
$\dfrac{\text{chest circumference}}{\text{trunk length}} \times 100$	104	152	187	165	217	160

Data from Schultz (1968).

Body proportions

Some exact body measurements of adult male gorillas are given in Table 3. Taking all three subspecies together, the average height is 172 cm, girth of the chest 147.3 cm and armspan 240.2 cm. The two eastern subspecies tend to be taller than the western lowland gorillas and to have larger chests. Mountain gorillas, however, have the shortest arms of any subspecies and hence have the shortest armspan. Some figures published by Eric Ashton, in 1954, indicate that young adult female *G. g. gorilla* and older adults, having well-worn teeth, average 136.1 cm in height, 112.4 cm in girth and have an armspan of 197.4 cm. Ashton's measurements for adult males seem to be rather low compared to those published by other authors. If we compare his data on females with those on males in Table 3, however, it transpires that female western gorillas are 81% as tall as males, with a girth 79% and armspan 84% of the male values.

The proportions of the gorilla's body yield much information about its evolutionary affinities and mode of life. Like the other apes and man, but unlike monkeys, gorillas have comparatively broad, shallow

Figure 14. Adult male western gorilla standing bipedally and beating its chest. Author's drawing from a photograph by R.D. Nadler.

chests, a short trunk and long arms (Figs 14, 15, 16). These features are made clear in Table 4 which compares various body proportions in macaques, apes and man. For instance, all the apes have longer arms than legs, particularly the gibbon and orang-utan which spend most of their lives in the trees. Hence the intermembral index, which expresses arm length as a percentage of leg length, is 165 in the gibbon and 172 in the orang, mainly due to elongation of their forearms. In the more terrestrial gorilla and chimpanzee, however, the indices are

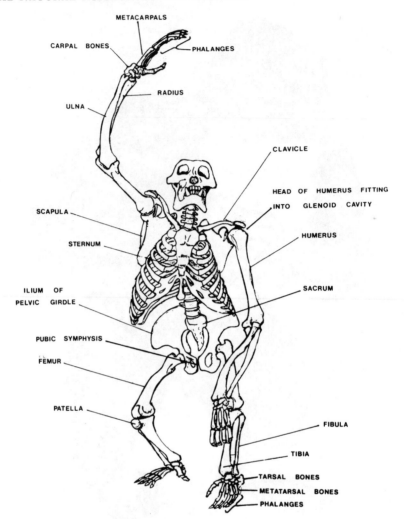

Figure 15. Mounted skeleton of an adult male western gorilla. After Raven (1950) with labelling added.

138 and 136 respectively whilst man has an intermembral index of 80.

The ancestors of the modern apes and man used their arms in a variety of ways to hang or swing from branches. Selection, therefore, favoured the development of longer arms, and a shorter, broader thorax with dorsally situated shoulder blades. This contrasts with the situation in macaques and most other monkeys which have the narrow, deep chests and laterally situated shoulder blades of typical quadrupeds.

Although many of the proportions of the gorilla's body attest to its

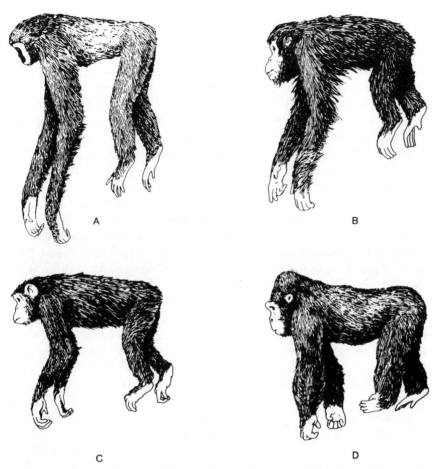

Figure 16. Side views of anthropoid apes, reduced to the same trunk length.
 A Gibbon **B** Orang-utan **C** Chimpanzee **D** Gorilla
After Erikson (1963).

tree-living ancestry it is now primarily a ground-dwelling ape and
has evolved various adaptations in this regard. Gorillas, like chimpanzees, are 'knuckle-walkers' and rest their weight on the backs of
the middle joints of the second to fifth fingers (Fig. 12). The skin in
these areas is greatly thickened to form callouses, and the hand and
wrist are otherwise modified for knuckle-walking. The gorilla's hands
are short and broad with a short but opposable thumb. Gorillas are
capable of making precision movements of the fingers; for instance
during feeding, which involves complex manipulations to select particular parts of plants. In the orang-utan, by contrast, the fingers are
greatly elongated so that the animal can grip or hang from branches.
When orangs walk on the ground, a rare occurrence except in zoos,

39

they often place their palms flat or clench their fists; hence the term 'fist-walking'. Knuckle-walking postures have been observed in one captive adult male orang, however, and Tuttle and Beck argue that since this highly arboreal ape may occasionally use such postures it is possible that the arboreal ancestors of the gorilla and chimpanzee had a predisposition to knuckle-walk when they descended to the ground. Since man is closely related to the African apes, the question naturally arises as to whether our ancestors evolved through a knuckle-walking phase. Some physical anthropologists argue that this was the case, whereas others, including Basmajian and Tuttle, believe that knuckle-walking did not occur in human ancestors. These workers at Yerkes Primate Center have made a unique study of the activity of various arm muscles during locomotion in the gorilla. To do this they implanted electrodes into particular muscles and recorded from them; a technique known as electromyography. They found that when a gorilla is standing in a knuckle-walking posture, there is in fact very little electrical activity in the long muscles which flex the fingers. Also, although there must be strong pressure tending to bend the hand dorsally, there is very little electrical activity in the muscles which flex the wrist in the opposite direction. They suggest therefore that it is the close packing of the wrist bones and positions of the ligaments in the wrist which maintain the hand in its characteristic position during knuckle-walking. Orang-utans show much greater wrist flexibility than the African apes do. Tuttle found that, in anaesthetized orangs, he could, on average, bend the wrists dorsally 85° as compared to only 30° in chimpanzees.

The legs and feet of gorillas also show some important adaptations for ground dwelling. Apes are able to grasp with their feet and this ability is best developed in the gibbon and orang-utan. Gorillas have an opposable big toe (Fig. 17). There is a good example of its use in Fig. 49 which shows a baby gorilla hanging by its feet. Adult gorillas most often use the feet for walking on the ground, however, and their big toes are closer to, and more in line with the other digits than in any other ape. This adaptation is most obvious in the mountain gorilla which also has the shortest arms and is the most terrestrial of the three subspecies. Both the African apes and man have thicker and more powerful big toes than the Asian apes do. It is also the case that the talus (foot bone) is relatively larger in the gorilla (where it forms forty per cent of the length of the entire foot skeleton), in man (fifty per cent) and in the chimpanzee (thirty-four per cent) than in the orang or gibbon, where it equals between twenty-six per cent and twenty-eight per cent. These values are taken from the work of Adolph Schultz and again indicate that the gorilla's feet are modified

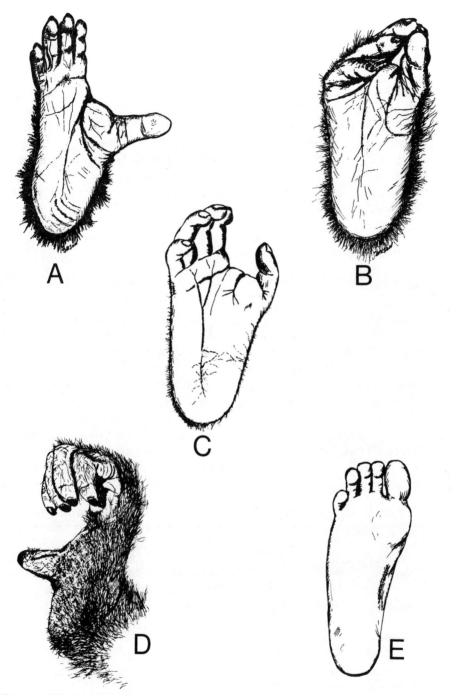

Figure 17. The feet of the great apes and man.
   A and **B** Infant western gorilla (8 months) showing the opposability of the great toe; **C** Chimpanzee; **D** Orang-utan. This specimen (a female) exhibits congenital absence of the terminal joint of the big toe; **E** Man. Authors drawings: not to scale.

for walking on the ground as are those of man and, to a lesser extent, the chimpanzee.

It is useful to compare the drawings of gorilla, chimpanzee and human foot with that of an orang-utan (Fig. 17). In the orang-utan the big toe is much reduced in size, and the terminal joint and toe nail are often missing. The flexor muscles of the orang's big toe are comparatively weak and when this ape climbs it mainly uses the very long second to fifth toes to grip the branches. Unlike the gorilla or chimpanzee, orangs have evolved as almost totally tree-dwelling apes and hence the proportions of their limbs, hands and feet are quite different from those of the African apes.

Structure and function of the skeleton

The skull

In adult gorillas the face is massive, with protruding jaws and a flattened nose. The eyes are overhung by prominent brow ridges and the cranium is comparatively small. If we examine Fig. 18 which also shows the head of a ten-week-old infant gorilla, then it looks quite different, for the face is small and flattened whilst the cranium is comparatively large. In order to appreciate the reasons for these differences it is necessary to consider the structure of the gorilla's skull and how it alters during growth in both sexes.

The skull consists of many bones fused together to form a hollow block. It has the dual role of protecting and supporting the brain and special sense organs and of forming the jaw apparatus. The cranium of a baby gorilla is comparatively large (Fig. 19). Schultz calculated

Table 5. Cranial capacities (in cc) of the apes and man

		Mean	Range	Cranial capacity as a percentage of body weight
Man	male	1400	1100–1700	2.1
	female	1300	1000–1600	2.2
Gorilla	male	535	420–752	0.3
	female	458	340–595	0.5
Chimpanzee	male	396	322–500	0.9
	female	355	275–455	0.9
Orang-utan	male	424	320–540	0.6
	female	366	276–494	1.0
Gibbon	male	104	89–125	1.9
	female	101	82–116	2.0

Data: from Schultz (1968).

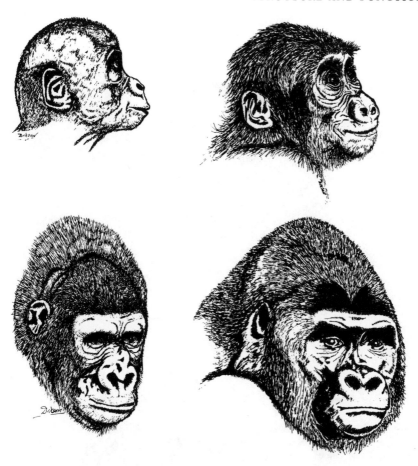

Figure 18. Portraits of infant and adult western gorillas. (*Upper left*) infant male aged 10 weeks; (*upper right*) infant female aged 8 months; (*lower left*) adult female; (*lower right*) adult male. Author's drawings: not to scale.

that the cranial capacity, which is almost equivalent to brain volume, averages 403 cc in baby gorillas weighing between four and twenty-five kilogrammes. Since in adult gorillas cranial capacities range from 420 to 752 cc in males and 340 to 595 cc in females, it follows that some baby gorillas actually have larger brains than certain female adults. Although the gorilla has in absolute terms the largest brain of any ape it has the smallest brain relative to its body weight (Table 5). The brain grows so little that in adult gorillas it is equivalent to only 0.3% of total body weight in males and 0.5% in females. This contrasts with the situation found in man, where a newborn infant's brain is only twenty-five per cent of the adult size and where adult brain weight equals more than two per cent of total body weight.

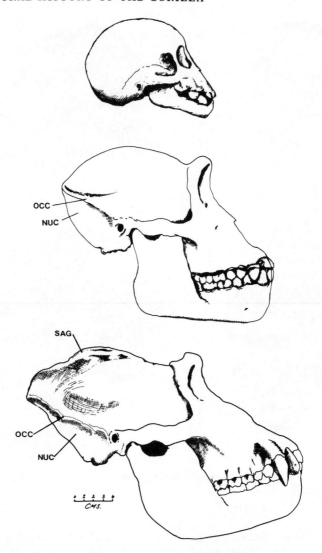

Figure 19. Skulls of infant and adult western gorillas. (*Upper*) infant; (*middle*) adult female; (*lower*) adult male. Sag. sagittal crest; Occ. occipital crest; Nuc. nuchal area. Author's drawings: to same scale.

Man is characterized by a striking postnatal growth of the cranium and brain whereas in the great apes it is the facial portion of the skull which grows dramatically after birth. If we examine a longitudinal section through the middle of an infant gorilla's skull (Fig. 20) it is clear that the orbits are situated well underneath the cranium, as is also the case in human infants. In man the orbits remain in this position throughout life, whereas in the gorilla they move forward

44

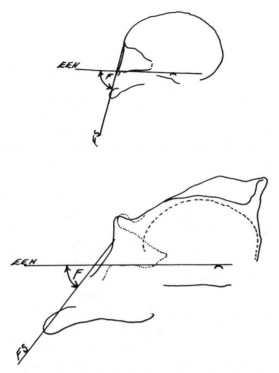

Figure 20. Sagittal sections through the skulls of an infant, and adult, male gorilla. (*Upper*) infant; (*lower*) adult male. EEH ear to eye horizon; FS facial slope; F facial angle. Redrawn from Schultz (1950).

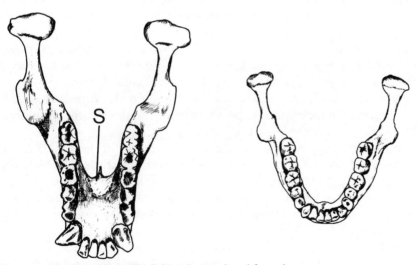

Figure 21. Mandibles of a gorilla (left) and man, viewed from above. **S** simian shelf. Author's drawings: not to scale.

45

and in adulthood lie well in front of the brain. A well-marked constriction therefore develops behind the orbits, and this is most pronounced in adult males. In both sexes prominent brow ridges develop, giving the skull a square-cut appearance around the orbits. The height of the facial region increases during growth in the gorilla, and there is a marked forward growth of the jaws. In the infant, the angle between the facial slope and the ear to eye horizon is almost 90°. The infant's face is almost flat. In the adult male, however, with its massive, prognathic face the facial angle is much reduced, though it shows a good deal of variability between individuals. The lower jaw of the adult gorilla is also massive (Figs 19 and 21). In side view it is apparent that the gorilla's lower jaw is not equipped with a projecting 'chin' as the human mandible is and which perhaps serves in man to strengthen the union between the two halves of the mandible. If the gorilla's mandible is viewed from above, however, then a strengthening bony shelf, the so-called 'simian shelf', is visible along the inner margin at the front of the jaw.

The gorilla's jaws require powerful muscles to operate them and the cranium provides only a relatively small surface area for the attachment of such muscles. In adult males there is a large crest of bone down the centre of the skull, called the sagittal crest. This crest only begins to develop when a male reaches adulthood and is largest in old silverbacks. It has the important function of providing additional surface area for the attachment of the temporal muscles (Fig. 22). It is possible that the crest may also serve to strengthen the sagittal suture against the opposing forces which operate upon it when the gorilla is chewing. Female gorillas have smaller jaws than males and lack a sagittal crest. As with some other features of the skull, such as the cranial capacity or facial angle, sagittal crests are very variable in size. In male orang-utans, which, like gorillas, have large jaws in relation to the size of the cranium, sagittal crests are commonly found, but they are uncommon in male chimpanzees and very rare in male gibbons. Some other primates, such as colobus monkeys, sometimes possess a sagittal crest.

Another feature which changes during growth in the gorilla is the position of the foramen magnum (the aperture through which the spinal cord passes) and of the occipital condyles, which articulate the skull with the vertebral column. In infants of all the catarrhine primates the foramen and occipital condyles lie underneath the cranium rather than at its posterior end as in other mammals. Man is unique in that the joint between the skull and the vertebral column remains under the cranium throughout life and hence the skull is balanced in an upright position on the neck vertebrae (Fig. 23). In

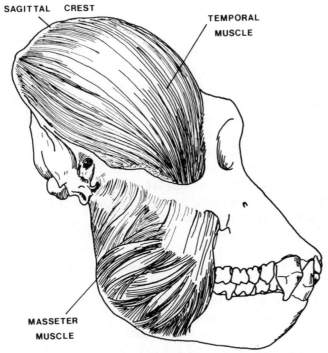

Figure 22. Skull of an adult male western gorilla, showing the positions of the temporal and masseter muscles. Redrawn from Raven (1950).

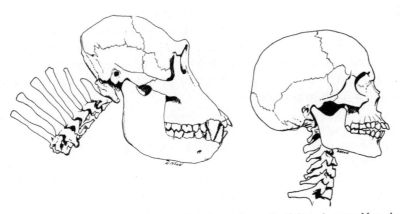

Figure 23. Lateral view of the head and neck skeleton of a gorilla (*left*) and a man. Note the elongated spines on the cervical vertebrae in *Gorilla*. Author's drawings.

47

the gorilla, however, the facial part of the skull enlarges considerably during growth so that in adulthood the condyles lie more posteriorly and the skull must be held in position by massive neck muscles. These muscles attach to the very large sloping nuchal area at the back of the gorilla's skull and to specially developed occipital crests. The neck vertebrae of the gorilla also show a special adaptation for the attachment of the neck muscles; their dorsal spines are greatly elongated. The great size of these spines and of the neck muscles limits the lateral mobility of this region so that a gorilla must turn the whole front of its body in order to look over its shoulder. It is appropriate to mention here that although the skull of the adult gorilla is massive and requires extensive supporting musculature, it is not a uniformly solid structure. The skull base and facial region contain an extensive system of sinuses or cavities, particularly in old males, and these probably serve to reduce its weight.

We have not yet dealt with the dentition of the gorilla, but a brief description should suffice. According to Schultz the deciduous or 'milk' teeth of the gorilla have all erupted by early in the second year. The incisors appear first, followed by the molar teeth and finally the canines. It seems probable, however, that gorillas do not complete the permanent dentition until they are ten or eleven years of age. Adults have thirty-two teeth which is the number found in all the Old World monkeys, apes and man. This can be expressed by the dental formula which gives the numbers and types of teeth in one half of the upper and lower jaws. For the gorilla, the formula is: $I\frac{2}{2}$ $C\frac{1}{1}$ $P\frac{2}{2}$ $M\frac{3}{3}$, where I = incisor, C = canine, P = premolar and M = molar teeth. The gorilla has an elongated palate and dental arcade, in contrast to the more rounded contour of the human tooth row (Fig. 21). Gorillas, like many primates but unlike man, have very large projecting canine teeth. Male gorillas have much larger canines than females, as can be seen in Fig. 19. It is also noteworthy that gorillas have very large molar teeth, presumably to deal with their tough, fibrous vegetable diet.

The vertebral column and rib cage

The rib cage of the gorilla is funnel shaped, wide at the bottom and narrow at the top (see Fig. 15). There are thirteen pairs of ribs and these are bent at their dorsal ends much more than the ribs of monkeys are. The same is true in all hominoids, for the ribs form a wide, shallow cage in contrast to the narrow, deep rib cage of monkeys. The first nine ribs are connected ventrally to the central sternum or 'breast bone'. In all the apes and in man the sternum is broad and consists of a number of bones which fuse together gradually with age. In monkeys, however, the sternal bones remain separate throughout life.

The gorilla has thirteen thoracic vertebrae, each of which bears one pair of ribs, and the neck consists of seven cervical vertebrae. The vertebral column is relatively shorter in hominoids than in monkeys but the shortening has not occurred in the neck and chest region. It is the lumbar region of the column which has become shortened during hominoid evolution. Whereas macaque monkeys have 7 vertebrae in the lumbar region, gorillas and chimpanzees have 3.6 on average, orangs have 4 and the gibbon and man have 5 lumbar vertebrae. Vertebrae from the lumbar region have become incorporated into the sacral portion of the column in apes and man. In

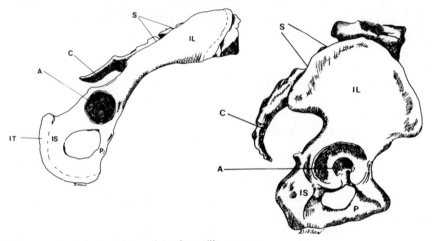

Figure 24. Lateral view of the pelvis of a gorilla and a man.
A acetabulum; **C** coccyx; **IL** ilium; **IS** ischium; **IT** ischial tuberosity; **P** pubis; **S** sacro iliac joint. Author's drawings: not to scale.

macaques, 3 vertebrae are fused to form the sacrum, whereas in gorillas there are usually 5 or 6. Beyond the sacrum there are a few vestigial vertebrae which represent the remnants of a tail. In all the apes and in man these vestiges have become fused to form the coccyx (Fig. 24). In man the sacrum is very wide and is bent dorsally at an angle of 60° to the long axis of the vertebral column. Part of the function of these adaptations is to ensure that the birth canal is wide enough to admit the large head of a human baby. In the gorilla the sacrum is not so broad as in man and it is bent dorsally at about a 30° angle. The birth canal is quite adequate, however, for the relatively small head of a baby gorilla to pass through.

The gorilla's spine, like that of other quadrupeds, acts mainly as a carrying beam, whereas in man it serves as a central supporting column during upright walking. Man shows certain modifications of the spinal column which make an interesting contrast with the gorilla. There is a pronounced ventral curve in the lumbar region of the human vertebral column and the lumbar vertebrae are thickened. These modifications render the column more springy and make it more resistant to the stresses which it must bear during walking. The lumbar region of the gorilla is shorter than in man and there is little indication of any ventral curvature. Nor are the vertebrae thickened. Duckworth once made the comment that: 'the vertebral column of the gorilla would be inadequate to the task of supporting the weight of the head and upper limbs, were not the latter still employed as supports and thus as a means of relieving the strain borne entirely by the vertebral column in the human type.' In fact, gorillas are quite able to stand erect and walk for short distances without using their arms as supports, although they do so comparatively rarely.

The limb girdles

It is worthwhile to consider functional aspects of the gorilla's pelvic and pectoral girdles. The pelvis (Figs 15 and 24) consists of two innominate bones, each of which consists in turn of three fused bones (ilium, ischium and pubis). The ilia are joined to the sacral portion of the spinal column dorsally, whilst the pubic bones meet ventrally to form the pubic symphisis and complete the solid loop of bone. The ischia form tuberosities which underlie the region upon which the gorilla sits. There are no specialized sitting pads or ischial callosities in gorillas, such as are found in the lesser apes and most Old World monkeys. The skin in the ischial region may be thickened in some gorillas but these are not true ischial callosities. Washburn, Rose and others have argued that callosities developed in Old World monkeys and lesser apes as adaptations for sitting in trees during feeding or sleeping. The great apes, particularly the terrestrial gorilla, employ different body postures during feeding, and they construct nests to sleep in. These facts perhaps explain why gorillas, chimpanzees and orangs lack ischial callosities, although ancestral forms may have possessed them.

The ilia of the gorilla's pelvic girdle are broad and very long so that they reach almost as far as the last pair of ribs. Because of this, lateral movements of the lumbar region are very restricted, as is also the case in the neck. The gorilla's trunk lacks the lateral flexibility found in the human skeleton. In the gorilla, the ilia act as levers transmitting the weight of the rear portion of the body from the sacro-iliac joint back to the hip joints. In man the situation is quite different, for the ilia are very short and broad and weight is transmitted downwards from the sacro-iliac region to the hip joints situated close by (Fig. 24). Since man walks upright all his weight is borne by the legs and Schultz calculates that the sacro-iliac joint has twice the area, per kilogramme of body weight, than in the quadrupedal gorilla. For the same reasons, man has relatively larger hip joints than the gorilla does.

The pectoral girdle invites close inspection because it provides a useful key to problems of primate locomotion and evolution. The pectoral girdle consists of the scapulae, or shoulder blades, which are the main mechanical members and the clavicles, or collar bones, which act as compression struts stabilizing the movement of the scapulae. The gorilla has long clavicles as do all hominoids, particularly the tree dwelling Asiatic apes; indeed they are twice as long relative to trunk length as the clavicles of monkeys. Unlike the apes,

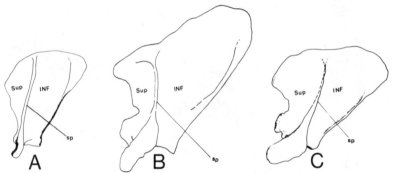

Figure 25. Scapulae of a macaque, an orang-utan and a gorilla, drawn so that the scapular spine is approximately the same length in each case.
 A Macaque: a quadrupedal monkey; B Orang-utan: a quadrumanous climber and arm swinger; C Western gorilla: a knuckle walker. Sup. supraspinous fossa; Sp. scapular spine; INF. infraspinous fossa. Author's drawings.

man's collar bones are positioned horizontally and this probably correlates with the fact that man walks upright and habitually holds and uses his arms below shoulder level.

The scapula is a roughly triangular plate of bone with a prominent ridge or spine along its outer surface. This spine divides the outer surface into the so-called supraspinous and infraspinous fossae (Fig. 25). The shoulder joint is situated at the narrow end of the scapula, where a cavity, the glenoid fossa, receives the head of the humerus. Unlike the pelvis, the scapula is not attached by bony joints to the vertebral column but instead is embedded in and moved by a complex set of muscles. Four of these muscles constitute the 'rotator cuff' which serves to stabilize the shoulder joint and control certain arm movements. The rotator cuff muscles tend to be well developed in all quadrumanous or arm-swinging primates due to the great stresses placed on their shoulder joints. Although the gorilla is mainly a ground-dwelling knuckle-walker, yet it has an arboreal ancestry and retains a broad, shallow rib cage and dorsally situated scapulae, as orangs and gibbons do. It follows that the stresses placed upon the gorilla's shoulder joint during knuckle-walking are not only different from those experienced by a climber such as the orang but differ also from those in a typical ground-dwelling monkey such as a macaque or baboon (Fig. 26). In a baboon, for instance, the laterally situated scapulae and the bones of the arm lie in the same mechanically advantageous perpendicular plane. In the gorilla, however, there are mechanically disadvantageous shearing stresses between the dorsal

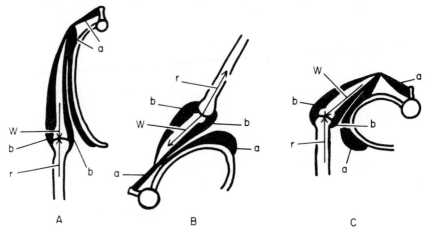

A   B   C

Figure 26. Potential stresses acting across the shoulder joint in various primates.
A Macaque: a quadrupedal monkey; **B** Orang-utan: a quadrumanous climber and arm-swinger; **C** Gorilla: a knuckle-walker; **W** body weight, **r** substrate reaction to body weight, **a** serratus and rhomboideus muscles, **b** subscapularis muscle. After Roberts (1975).

scapula and the forearm at the shoulder joint. It is perhaps for this reason that the gorilla has such a massive, broad scapula which is specially strengthened by secondary lines of thickening.

Some aspects of the internal organs

The respiratory, vascular, alimentary and urinogenital systems of the apes and man show considerable uniformity in their general structure. There are, of course, departures from the common pattern, the phylogenetic significance of which is not always clear as in the case of skeletal variations. For example, the liver of the gorilla shows considerable sub-division into lobes and resembles the liver of monkeys much more than other hominoids. Flower, in 1872, was the first to comment on the dilemma: 'either that modifications of the liver are not very characteristic in natural and related groups of animals, or that the gorilla ought not to occupy that system which has hitherto been accorded to it.'

However, so many structural characteristics of the gorilla indicate its close relationship to the other apes and to man, that not much importance can be attached to an isolated observation on its liver. When assigning evolutionary relationships it is best to compare many characters and not to rely upon comparisons of a single organ or system. I intend to select only a few structures which are of interest from a functional point of view. Detailed accounts of the gorilla's internal organs are available in the *Raven Memorial Volume* or in Duckworth's *Morphology and Anthropology*, Vol. I.

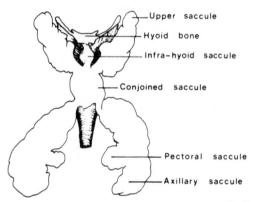

Figure 27. Diagram of the laryngeal sac of an adult male western gorilla. Redrawn from Duckworth (1915).

An important and distinctive feature of the respiratory system of the gorilla is the great extension of its larynx to form air pouches (Fig. 27). The sacs emerge from each side of the larynx, near the top, just below the hyoid bone which lies at the base of the tongue. The two sacs merge to form a 'conjoined saccule' lying over the front of the larynx, but not before they have each given off an upper saccule which extends anteriorly and backwards into the neck. The largest part of the system extends posteriorly and outwards from each side of the conjoined saccule. These posterior diverticula pass between the outer muscles of the chest wall and rib cage, and reach as far as the armpits.

The air pouches are much more extensive in adult males than in females or young gorillas. It seems very likely that the sound a male makes by beating his chest is enhanced by inflating the laryngeal sacs.

A hollow 'pock-pock' note is produced by silverback males in contrast to the dull thumping chest beats of females or young animals.

Laryngeal pouches are also found in the orang-utan, chimpanzee and siamang, but not in the gibbon. In man, two small crypts in the larynx represent the remnants of laryngeal pouches. It seems reasonable to assume that laryngeal pouches evolved to serve as resonators or perhaps as reservoirs of air for prolonging certain vocalizations. Male *Cercopithecus* monkeys possess larger laryngeal pouches than females do and inflate them during territorial vocal displays. The same is true of the enormous laryngeal diverticula of the male orang and of the pouches which occur in both sexes of siamang. It is puzzling, therefore, that gibbons, which lack laryngeal pouches, produce vocalizations just as impressive as those of siamangs.

It is worthwhile to consider the structure of the gut in gorillas, because they consume very large amounts of vegetation and one might expect to find some specialized structural adaptations. Actually, there are surprisingly few differences between gorillas and man in this respect. The gorilla has a very large stomach, for obvious reasons, but there is nothing approaching the complex three-chambered arrangement found in the leaf-eating *Colobus* monkeys. The small bowel (duodenum, jejunum and ileum) measures about ten metres in the gorilla and about five metres in man but, apart from size, the small and large bowels are very similar in the two species. The gorilla has a large caecum, and there is a vermiform appendix as in man. It is possible that the gut of the gorilla contains symbiotic micro-organisms which break down cellulose and assist the animal to digest plant food. Symbiotic protozoa are found in ruminants such as cows, and Ohwaki and his co-workers have shown that bacterial fermentation occurs in one species of *Colobus* monkey (*C. polykomos*). Hooton states that, according to Reichenow: 'The intestines of wild gorillas are inhabited by a species of infusoria which digest cellulose and, consequently, help their hosts to make use of plant foods. . . . these infusoria disappear in a few weeks from the intestines of captive animals.' Unfortunately there appears to be no other literature on this subject, but if symbiotic micro-organisms do occur in the intestinal tract, it might explain how gorillas are able to digest their high cellulose diet.

In Chapter 7, I shall be dealing with reproduction and so it is necessary to say a few words about the structure of the gorilla's reproductive organs. The external genitalia of an infant and an adult male are shown in Fig. 28. In the newborn, the testes have descended but are often retracted by a cremasteric reflex and so are not visible in the scrotum. The scrotum is very small, in contrast to human

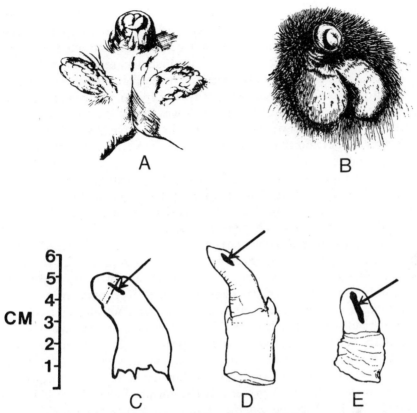

Figure 28. The external genitalia of male gorillas and other great apes.
A infant western gorilla (enlarged); **B** adult western gorilla (reduced); **C** penis of adult western gorilla to show position of baculum (arrowed); **D** penis of adult chimpanzee to show position of baculum (arrowed); **E** penis of Bornean orang-utan to show position of baculum (arrowed). C, D, and E are scale drawings from radiographs. **A** drawn from Carter (1973). **B–E** , Authors specimens and drawings.

newborns, in which it is large and in which the testes descend before birth. In the adult gorilla, the external genitalia are very inconspicuous and obscured by the surrounding hair. The scrotum is bilobed and the testes are small and bean shaped. Wislocki recorded that the testes of one adult western lowland gorilla each weighed eighteen grammes and Raven's observations on another male indicate that each testis measures about 4×2 centimetres. Hall-Craggs found that a testis from an adult mountain gorilla measured 5.5×3.3 centimetres and weighed 20.6 grammes (including the epididymis). It is important to note that these figures all refer to wild-shot specimens, since, in captivity, some gorillas have much smaller testes and testicular atrophy has been documented in seven specimens.

Raven gives the length of the penis in the gorilla as 9 centimetres, but I think this must refer to its length in dissection and not to the external dimensions. Even when erect the organ probably measures less than 7 centimetres, although, not surprisingly, no one has made an exact estimate. Warren Thomas, for instance, mentions 14 centimetres but this is presumably a guess. The gorilla, like other primates (exept for instance the woolly monkey, the spider monkey and man), possesses a penile bone or baculum. It was described by Osman Hill and Harrison Matthews, who showed that it is very small as in the other great apes and that it may represent the vestiges of a previously larger structure.

Male chimpanzees have much more prominent external genitalia than gorillas do, and we shall see in later chapters that these two apes differ in many aspects of their sexual and social behaviour. It is also the case that the female chimpanzee has a 'sexual skin' which swells during the first half of the menstrual cycle (see Plate 20). Lengthening of the penis in the chimpanzee may have occurred during evolution for purely mechanical reasons, since the female's swelling adds considerably to the depth of her genital tract. By contrast, the external genitalia of adult female gorillas are inconspicuous and consist of a single small pair of labia. In the human female there are two pairs of labia; the labia minora which derive embryologically from the urethral folds and a larger outer pair of labia majora which develop from the genital swellings in the foetus. Authorities differ in their opinions of which pair of labia is present in the gorilla. Raven records that a young female which he examined possessed only labia majora but Atkinson and Elftman state that labia majora are only present in the foetus and that the structures present in adult females are labia minora. These details need not concern us but I mention them because labial size is an important indicator of the stage of the menstrual cycle in gorillas. The labia swell slightly at mid-cycle and are smallest during menstruation (see Plate 20).

The internal genitalia of the female gorilla are similar to those of other hominoids. The uterus and vagina lie almost in line with each other, whereas in woman the uterus is inclined ventrally at a sharp angle to the vaginal canal. Some of the differences in the arrangement of the internal genitalia may simply reflect the fact that the pelvic cavity of the gorilla is shaped quite differently to that of woman.

The gorilla has large ovaries, four centimetres long in one specimen dissected by Elftman and Atkinson. Each ovary has been estimated to contain 53,000 oocytes as compared with 225,000 in the orangutan and 154,000 in the chimpanzee. These data refer to very few specimens. Only one gorilla and one orang-utan were studied, and since in human females between 100,000 and 200,000 oocytes occur in each ovary, it is probable that apes also exhibit much individual variation. The ovaries are also endocrine glands, secreting mainly oestrogens and progesterone. Rhythmic changes in secretion of these hormones bring about changes in labial swelling, the uterine lining and vaginal epithelium during each menstrual cycle. The hormonal control of sexual behaviour in gorillas is dealt with in Chapter 7.

Chapter 4

How Close to Man?

IN his *Memoir on the Gorilla* of 1865, Richard Owen listed the changes he considered necessary to 'transmute a gorilla into a man'. The list is short and involves changes in the anatomy of the brain, teeth and intestines. Obviously this understates the gulf which separates the gorilla and man, but an examination of the relationship between the two forms is very worthwhile.

The anatomical similarities between man and gorilla have already been discussed and it is clear that the two forms share a common ancestor, but how long ago did this creature live? How closely is the gorilla related to the chimpanzee and the orang-utan or to the gibbon and siamang? My review of these questions must inevitably be over-simplified because of the diversity of opinions on the subject and because there are not sufficient fossils to provide perfect answers. For some mammalian groups, such as the horses, there is an extensive fossil record which provides a firm basis for comparing the extant forms. Unfortunately this is not the case where the apes and man are concerned. In considering possible evolutionary relationships between the apes and man, however, three theories may be formulated. Ernst Mayr has summarized these as follows:

1. 'The hominid line [the line leading to man] branched off from the stem of the living anthropoids before they split into three separate lines.'
2. 'The hominid line branched off well after the gibbon line but before the pongids split into the lines which gave rise to *Pan*, [*Gorilla*] and *Pongo*.'
3. 'The hominid line branched off from the line of the African apes at a comparatively recent date, long after the pongid line had split into an Asiatic and African branch.'

The first of these theories, depicted in Fig. 29, may be discarded. Fossil gibbons, belonging to the genus *Pliopithecus*, have been found

58

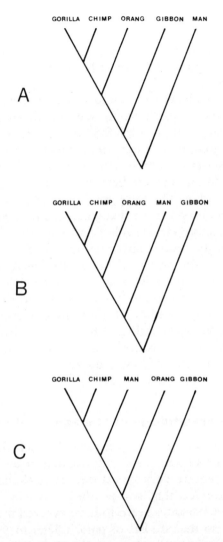

Figure 29. Three schemes (**A, B, C,**) depicting possible evolutionary relationships between the apes and man. The lengths of the branches and distances between nodes are arbitrary and do not represent a time scale.

in deposits about eleven million years old in India and also in Europe dating back sixteen million years. *Limnopithecus*, a closely related ape, lived in Africa about fourteen to twenty-three million years ago. It would seem therefore that the ancestors of today's gibbon and sia-mang branched off at an early stage from the line which lead to the great apes and man. It is always possible, however, that *Pliopithecus* and *Limnopithecus* are not ancestral to the gibbon and siamang, but resemble them only because of adaptation to a similar ecological niche. As we shall see, however, comparative studies of the living apes support the argument for an early divergence of the forerunners of the lesser apes.

There are not sufficient fossils to prove which of the remaining two schemes is the correct one. At one time or another, authorities such as Sir Arthur Keith, Adolph Schultz and Henry Fairfax Osborn have proposed hominoid family trees which place the gorilla and chimpanzee closer to the orang-utan than to man. Others, including Sonntag, Elliot-Smith and Weinert, place the African apes closer to man than to the orang. The trees put forward by these workers differ in various details, however, so that whereas Elliot-Smith considered the gorilla to be more closely related to man than the chimpanzee, Weinert thought the reverse situation more likely.

We have seen in the previous chapter that many insights derived from the study of comparative anatomy indicate that the gorilla and chimpanzee may be more closely related to man than to the orang-utan. More recently, studies of the chromosomes, parasites and biochemistry of the apes and man have strengthened this view. It will be most useful to describe some of this research first before considering the problem of how long ago the common ancestors of gorilla, chimpanzee and man lived.

Comparative parasitology of the apes and man

In 1891, von Ihering first put forward the idea that studying the parasites of animals might yield information of evolutionary value. Many parasites require very specialized physiological conditions to survive and reproduce, that is to say they have a high degree of 'host specificity'. Some parasites seem to have evolved much more slowly than their hosts so that studies of parasitology may yield important information about how closely related the host species are to each other.

Dunn and Garnham have both reviewed information on the malaria parasites of the apes and man. Malaria is caused by uni-cellular organisms belonging to the genus *Plasmodium*. A great deal of host specificity occurs among malaria parasites, partly because each species is adapted to utilize the particular type of haemoglobin found in the blood of its host. However, it is important to mention here that host specificity is not an absolute rule among malaria parasites. Thus the Colombian owl monkey can act as a host for human *P. falciparum*, a finding which was first reported by Geiman and Meagher in 1967.

Despite its remote relationship with man, the owl monkey therefore plays a unique role in research on human malaria. Four species of malaria parasites are found in man (see Table 6) and one of them, *Plasmodium ovale*, has not been found in any other hominoid. The other three species, however, occur as very similar forms in both chimpanzees and gorillas, but not in the orang-utan or gibbon. *P. schwetzi* is found in both African apes and is very similar to human *P. vivax*. *P. reichenowi* of the African apes is very similar to the *P. falciparum* of man. Chimpanzees are also parasitized by *P. rodhaini*, a form almost indistinguishable from human *P. malariae*, but which has not been found yet in the gorilla. All the information on malaria

Table 6. Malaria parasites of man and the apes

	Plasmodium species
Man	*P. malariae*
	P. vivax
	P. falciparum
	P. ovale
Gorilla	*P. schwetzi*
	P. reichenowi
Chimpanzee	*P. schwetzi*
	P. reichenowi
	P. rodhaini
Orang-utan	*P. pitheci*
	P. silvaticum
Gibbon	*P. hylobati*
	P. youngi
	P. eylesi
	P. jefferyi

Data from Dunn (1966) and Garnham (1973).

parasites in gorillas refers to the western lowland sub-species; there have been few investigations of eastern gorillas and all have been negative.

The malaria parasites of orang-utans show some similarities to those of gibbons, but are different from the *Plasmodium* species found in the gorilla, chimpanzee or man. Garnham states that when a human volunteer was bitten by a mosquito infected with *P. pitheci* from an orang-utan, he failed to develop malaria.

Table 6 does not include information on the siamang, since, although malaria parasites have been found in these apes, the species of parasite was not determined. One last point of interest is that protozoan parasites of the genus *Hepatocystis* have been found in one of the gibbons (*H. concolor*) and also recently in the orang-utan by

Stafford and his colleagues. *Hepatocystis* is common in monkeys but has not been found in the African apes or in man.

Frederick Dunn has also made some interesting comparisons of the species of helminth worm parasites found in man and the apes. Some helminths have a low host specificity whereas others are restricted in their ability to parasitize potential hosts. Thirty-four genera of helminth worms have been reported in the various hominoids and seven of them are common to all the apes and man. If the remaining twenty-seven genera are considered, however, much overlap in the

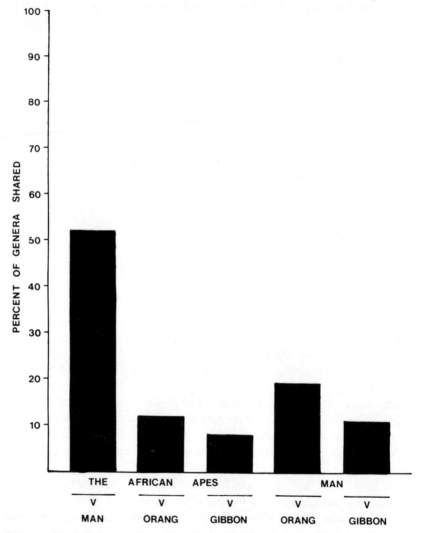

Figure 30. Helminth parasites shared by man and the apes. Based upon data in Dunn (1966).

occurrence of helminth species occurs between gorillas, chimpanzees and man. Less overlap is apparent between any of these forms and the orang-utan or gibbon. The African apes and man share eleven of twenty-one genera (fifty-two per cent) whereas the African apes share only three of twenty-two genera (fourteen per cent) with orang-utans and two of twenty-four genera (eight per cent) with gibbons (Fig. 30).

A word of caution is required about the orang-utan, since more research may be necessary to identify all its helminth worm parasites. The conclusion that gorillas, chimpanzees and man share a greater percentage of parasites seems justified, however, and this fact testifies to the close relationship between them. It is interesting, as well, that three genera of external parasitic arthropods; *Phthirus*, *Pediculus* and *Sarcoptes* are found on the African apes and on man, but not on the Asiatic gibbon and orang-utan.

Comparative studies of chromosomes

A comparison of the gorilla's chromosomes with those of other apes and man is most fruitful, for it is the chromosomes which contain the genetic material. Each cell nucleus contains a complete, or so-called 'diploid' set of chromosomes, half of which derive from the male parent and half from the female. Normally the chromosomes are not visible, but when a cell divides, by mitosis, the chromosomes can be observed. Each chromosome then consists of two ribbon-like structures of variable length, called chromatids, which are united by a small centromere. The centromere may be situated near the middle of the chromatids (the metacentric condition) or may unite them near their ends (the acrocentric condition). It is possible to culture white blood cells from apes or man and to arrest the process of cell division by adding colcemid to the culture medium. The chromosomes may then be counted and their morphological or staining properties may be studied. It is possible to arrange the chromosomes in numbered pairs, characteristic for the species, depending upon their shape, the position of the centromere and lengths of the chromosome arms, as well as the banding patterns produced by Giemsa, Quinacrine and other staining techniques.

The diploid number of man is 46 chromosomes, as was first recorded by Tijio and Levan in 1956. The diploid numbers for the apes are as follows: gibbon 44, siamang 50, orang-utan 48, chimpanzee 48 and gorilla 48. Since all the great apes share the same diploid number it might seem appropriate to conclude that they are more closely related to each other than to man. Chiarelli has stated this view as follows: 'The superfamily Hominoidea should be restricted

63

to the anthropoid apes (*Pongo, Gorilla, Pan*) and to man. Among these, man is closely distinguished by the number of chromosomes and must be classified in the family Hominidae; while the true apes will constitute the sub-family Pongidae.'

There can be no doubt, however, that the diploid number is a weak criterion upon which to argue evolutionary relationships, for diploid numbers are very variable even between closely related primate genera and species. There is no question, for instance, that the gibbon and siamang are very closely related, yet one has a diploid number of 44, as have the *Colobus* monkeys, and the other has 50 chromosomes in its karyotype. The proboscis monkey, *Nasalis*, has a diploid number of 48 just as the great apes do, but it is only remotely related to them. As a final example we can consider the guenon (*Cercopithecus*) monkeys, all very similar morphologically but whose karyotypes range from 54 to 72, and in certain guenons the diploid number varies within a single species. It follows, therefore, that differences or similarities in the diploid number between the apes and man are not sufficient grounds alone for classifying the three great apes together and separating them from man.

It is more important to consider the comparative anatomy and staining properties of chromosomes among the hominoids. The chromosomes of man, the gorilla, the chimpanzee and orang-utan are remarkably similar, but very different from those of gibbons. All the chromosomes of the gibbon (*Hylobates lar*) are metacentric, whereas both metacentric and acrocentric chromosomes are found in the great apes and man. Gibbons possess a pair of chromosomes with a deep constriction in one arm. These 'marked' chromosomes occur in Old World monkeys as well, but not in the great apes or man. By contrast, the chromosomes of the chimpanzee and man are difficult to tell apart during cell division unless they are counted. Thirteen chromosomes pairs are virtually identical; nine others differ slightly in positioning of the centromere. Sixteen chromosome pairs found in the gorilla are almost identical to those found in chimpanzees but differ from those of the orang-utan.

Chromosome banding patterns of the great apes and man are very similar. Millar points out that 'ninety-eight to ninety-nine per cent of the 500 or so bands observed with the general banding methods are homologous in these four species.' More detailed studies have revealed some important differences, however. The orang-utan has fewer banding patterns similar to those of man than does the gorilla or chimpanzee. The fluorescent dye quinacrine produces certain 'Q brilliant' patterns on the chromosomes of man and the African apes which do not occur in the orang-utan. In Millar's opinion, these detailed studies also point to a very close relationship between the gorilla and man. Thus, apart from man, the gorilla is the only mammal which exhibits Q brilliant staining of its Y chromosome.

The banding studies also provide clues about evolutionary changes in the chromosomes of the apes and man. Chromatids may sometimes fragment and rejoin in various ways during the cell divisions necessary to produce ova and spermatozoa. Such 'inversions' and 'transloca-tions', as they are called, are mutations. It is thought that a number of such chromosomal re-arrangements have occurred during evolu-tion of the great apes and man, and that many of the differences between them result from such mutations.

Biochemical studies

Since 1904, when Nuttall published his classic study entitled *Blood Immunity and Blood Relationship*, it has been known that human blood proteins resemble those of apes very closely. Modern research is based upon this fact and much of the biochemical evidence regard-ing the phylogenetic relationships of apes and man derives from studies of blood chemistry.

One way of comparing blood proteins is to investigate their immu-nological properties. For instance, if blood serum from a chimpanzee is injected into an Old World monkey then the monkey produces antibodies to the chimpanzee proteins. Antiserum taken from the treated monkey may then be tested against serum from the various apes and man, when the antibodies react with proteins (antigens) in the test serum by a 'precipitin reaction'. The usual method employed is to prepare slides coated in agar and place the antiserum and test serum in wells scooped out of the agar's surface. The antiserum and test serum then diffuse through the agar to produce 'precipitin lines' at the points where the antigens are at optimal concentration. The reaction may be hastened by applying an electric current to the preparation, a process called immunoelectrophoresis. The results of one such experiment are shown in Plate 7, in which antiserum raised in an Old World monkey by injecting it with chimpanzee serum was tested against sera from the various apes and man. The degree of similarity between the serum proteins of chimpanzees and other hominoids is indicated by the similarities of their precipitin reactions. Man and gorilla are very similar to the chimpanzee, whereas the orang-utan and gibbon are clearly differentiated. Many experiments of this kind have been carried out, by raising antisera to the different apes or man and then testing these in the way described. All experi-ments yield the same result: that the gorilla and chimpanzee are more similar to man than to the orang-utan or gibbon. Goodman has found that thirteen out of fourteen human serum proteins he has studied are more similar to those of the African apes than to those of Asiatic apes.

The similarities between the antigenic properties of these proteins reflect similarities of the sequencing of the amino acids from which they are constructed. This is because the antigenic sites which react with the antibodies are shaped by groups of amino acids situated on the surface of the protein. Other studies comparing proteins in apes and man have sought to determine the number and sequence of amino acids in the molecules. The most complete information has been obtained for man and chimpanzee; many proteins such as haemoglobin and myglobin have been shown to be virtually identical in the two species. Both haemoglobin and myoglobin differ in only one amino acid replacement in the chimpanzee as compared to man. It follows that the genes on the chromosomes, which encode information to produce these proteins, must also be virtually identical in the two species. King and Wilson argue that, based on protein studies, the genetic distance between chimpanzee and man 'is very small, corresponding to the genetic distance between sibling species of fruit flies or mammals'. This argument presumably also applies to the gorilla since it is so closely related to the chimpanzee. King and Wilson envisage that the vast differences between chimpanzee and man as regards anatomy, physiology and behaviour are not the result of mutations of single genes but rather of mutations such as translocations and chromosomal fusions, which have caused a re-positioning of certain genes on the chromosomes. This idea fits with the observations on the staining properties of chimpanzee and human chromosomes.

It is as well to remember that these arguments are based on work involving blood proteins. The genetic codes of blood proteins may be very similar in the African apes and man but this may not be the case for the genes which encode for other large molecules. The fact also remains that the important differences between the gorilla or chimpanzee and man are to be found by considering the anatomy, physiology and behaviour of whole animals rather than the structure of particular molecules in their blood.

The problem of time scales

The gorilla and chimpanzee are the apes most closely related to man, but how long ago did the common ancestors of these forms exist? The traditional view is that man's ancestors split off very early from the line which gave rise to the African apes. Fig. 31 shows a diagram of the Cenozoic, an era of geological time which began about sixty-five million years ago and which is divisible into six epochs; Paleocene, Eocene, Oligocene, Miocene, Pliocene and Pleistocene. Many have suggested that the hominid line split from the African ape line during the Miocene epoch, about fifteen to twenty million years ago. Gregory, Sonntag and Elliot-Smith have all proposed such a scheme. Others, including Osborn and Keith, placed the divergence earlier still, during the Oligocene.

In 1837, Lartet described the lower jaw of an anthropoid from

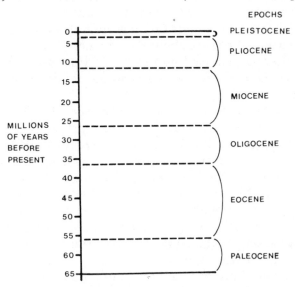

Figure 31. Time scale of the Cenozoic era. Redrawn and modified from Pilbeam (1972).

early Miocene deposits in France. This he named *Dryopithecus fontani*; and he regarded it as a possible hominoid ancestor. *Dryopithecus* remains have been discovered in Europe, Africa and Asia. Various authors have suggested that the ancestors of the gorilla, chimpanzee (and perhaps of man) belonged to this group. The fossil history of the orang-utan remains unknown, although Smith and Pilbeam have recently suggested that the ancestors of the orang-utan may have been terrestrial apes. Dryopithecine remains were found in India by Pilgrim. African forms were discovered in Kenyan and Ugandan Miocene deposits which date back eighteen to twenty-three million years. Some of these African finds were placed in the genus *Proconsul* but Simons and Pilbeam of Yale University suggest that *Proconsul* is so similar to *Dryopithecus* that the two genera should be united. *Dryopithecus major* was intermediate in size between the modern

67

chimpanzee and gorilla. Its remains have been found in areas where volcanoes occurred during the Miocene so it is possible that it lived in montane rain forests on volcanic slopes, as does the mountain gorilla today. Its foot-bone resembles that of the gorilla more than the chimpanzee and it is possible that it was a knuckle-walker, although Schon and Ziemer of Johns Hopkins University suggested that the closely related *D. africanus* was not a knuckle-walker. Such disagreements are commonplace in the extensive literature which deals with the scanty fossil evidence of ape evolution. David Pilbeam has suggested that *D. major* may have been ancestral to the gorilla, whilst its smaller relative, *D. africanus*, could represent a forerunner of the chimpanzee. He points out, however, that the fossil record is not yet sufficient to be sure about the affinities of the various dryopithecines. There are, in fact, no fossils to fill the fifteen-million-year gap between *D. major* or *D. africanus* and the modern gorilla and chimpanzee.

One more fossil genus must be mentioned briefly and that is *Ramapithecus*, a creature which possessed distinctly hominid facial and dental characteristics. It was discovered in India by Lewis in 1934, and it was this anthropologist who first suggested that *Ramapithecus* belonged to the line which led to man. *Ramapithecus* lived during the late Miocene and early Pliocene. Specimens found in Africa, originally classified as *Kenyapithecus wickeri* probably belonged to *Ramapithecus*. A possible human and possible gorilla and chimpanzee ancestors therefore existed during the Miocene, fifteen to twenty million years ago.

Many theories have been advanced to fill the vacuum left by the poor fossil record of ape and human origins. I hope that it will not confuse matters if I give two examples. The majority of anthropologists would probably agree that man has an African origin; most of the fossil evidence of hominid evolution has been located in Africa and comparative studies of gorillas, chimpanzees and man indicate their close relationship. However, Benveniste and Todaro have suggested an Asiatic origin for the hominid line, based on their genetic studies of Old World monkeys, apes and man. All these forms include, within their DNA, gene sequences (virogenes) related to the RNA of a virus isolated from baboons. Comparison of virogene sequences with other cellular gene sequences allowed Benveniste and Todaro to distinguish clearly between African apes and monkeys on one hand and Asiatic apes and monkeys on the other. Man fell with the Asiatic group, not the African one, however. These workers point out that since *Ramapithecus* occurred in Asia, as well as in Africa during the Miocene, it is possible that hominids stem from Asiatic ancestors.

Modifying this idea slightly at the suggestion of Dr R.D. Martin, their results may indicate that the hominid line underwent a substantial phase of its evolution in Asia, but not necessarily that it originated there.

Since there is no clear picture of what hominid and ape ancestors were like, Zihlman and co-workers have suggested that the pigmy chimpanzee might represent a 'prototype for the common ancestor of humans, chimpanzees and gorillas'. *Pan paniscus* is smaller and more gracile than the common chimpanzee, and exhibits less sexual dimorphism in canine tooth size and other features. The arms and legs of the pigmy chimpanzee are of roughly equal length, whereas the common chimpanzee has longer arms. On the face of it, the pigmy chimpanzee appears to be a less specialized animal embodying features which occur in certain early hominid fossils. These are interesting observations but the possibility must be considered that these are derived and not primitive features for *Pan paniscus*. The pigmy chimpanzee occurs only in a restricted area of the Congo (Zaire) basin (see Fig. 5). Is it possible that it developed from a larger ancestor under conditions of partial isolation? Tribes of pigmy humans are also found in the Congo basin and it may be that unusual zoogeographical conditions have influenced their evolution. This argument perhaps emphasizes that, although the extant apes are invaluable for comparative studies, it is not advisable to treat any one of them as a prototype ancestor.

In recent years biochemical studies have also provided information which conflicts with the fossil evidence of ape and human evolution. Sarich and Wilson of the University of Berkeley in California have studied the serum albumens of the apes and man. They calculated the 'distances' which separate the various species, basing their calculations of the time needed to produce such distances on the hypothesis that albumens have evolved at a constant rate in the various lines. Sarich and Wilson calculate that the gibbon line separated about ten million years ago, the orang-utan seven million years ago and that a three-way split about four million years ago produced lines leading to gorilla, chimpanzee and man.

It is clear that if *Dryopithecus* and *Ramapithecus* are ancestral to the African apes and man respectively, then the biochemical datings proposed by Sarich and Wilson cannot be correct. If, however, the dates based on biochemical information are accurate, then the Miocene fossils are not ancestral to gorillas, chimpanzees and man in the way discussed. The discrepancies between the fossil and biochemical schemes are shown in Fig. 32. What is most encouraging is that the relationships between the various apes and man are essentially the same in both trees; it is only the timing of the divergences which differs.

Some biochemists do not agree with Sarich and Wilson that blood proteins have evolved at the same rate in all the lines which led to today's Asiatic apes, African apes and man. Morris Goodman and William Moore, for instance, suggest that proteins such as haemoglobin may have changed more slowly during ape and human evolution than in other mammals. They suggest that the deceleration in

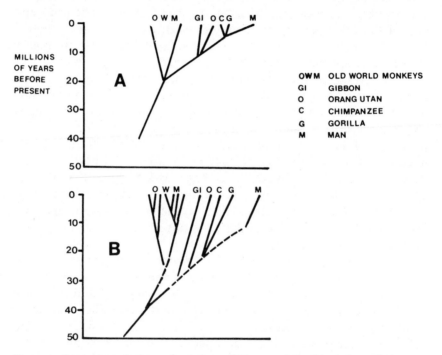

Figure 32. Comparison of schemes for dating evolutionary relationships between the apes and man. **A** based on biochemical evidence; **B** based upon fossil evidence. Redrawn and modified from Pilbeam (1972).

protein evolution occurred possibly because the hominoids have a much longer generation length and also a more complex internal organization than other mammals so that mutations resulting in changes in blood protein structure were more likely to prove disadvantageous. I present their viewpoint because if it is adopted, the biochemical evidence may be made to fit the available fossil record. I will not presume to evaluate the biochemical studies summarized above, since I lack the necessary knowledge to do so. It seems safe to assume, however, that the relationships between the apes and man portrayed in Figs 29c and 32 are correct. The gorilla's closest relative is the chimpanzee and both these African apes share man as a close relation. Whether a three-way split occurred between the lines which gave rise to the gorilla, chimpanzee and man remains to be seen. It may be that the hominid line diverged somewhat earlier from that which gave rise to the African apes. The orang-utan springs from a more ancient stock and the ancestors of the gibbon (and siamang) diverged earlier still. As to the dates of those divergences and the exact ancestors of the gorilla no certain answers can be given. Many of the theories summarized in this chapter may be discarded in future when more fossils are located.

Chapter 5

Senses and Intelligence

AT a quantative level we know very little about the gorilla's sense of sight, smell, hearing, touch or taste for the animal has rarely been studied to measure these attributes. Our only insights come from incidental observations made during field studies or work on captive gorillas. Some extrapolation is also possible from experiments on closely related forms, particularly the chimpanzee and orang-utan.

As regards the visual sense, gorillas are diurnal creatures and almost certainly possess excellent colour vison. Laboratory work has shown that gibbons, orang-utans and chimpanzees all have colour vision. A young female mountain gorilla tested by Yerkes was able to locate food in containers of various colours including green, yellow and red, as well as black and white. Jones and Sabater Pi also record that western lowland gorillas prefer the ripe red fruits of *Aframomum*, but other cues beside colour might be involved in making such a discrimination. Sabater Pi has observed that fifty-one per cent of foods eaten by the western lowland gorillas of Rio Muni vary in colour from green through greenish-yellow to pale yellow. Another thirty-eight per cent of foods vary from intense red to pale orange in colour. These are mostly fruits, and since young gorillas show intense interest when encountering trees bearing such fruits, and then climb up to investigate, it is possible that colour acts as a cue. The visual sense is undoubtedly of paramount importance to gorillas, being essential for navigation in the forest, for locating food and communicating with other group members. In the gorilla, as in all higher primates, we find the full expression of an evolutionary trend toward the development of stereoscopic vision and a greater reliance upon this sense than upon olfaction. Schaller reached the conclusion that, in the mountain gorilla, the sense of sight is probably roughly the same as in man.

Compared to the prosimian primates and New World monkeys, gorillas and other apes make little use of olfactory communication and their sense of smell is probably comparatively weak. Gorillas sometimes sniff food items before eating them and will treat novel objects in a similar fashion. Sabater Pi records that gorillas in Rio Muni reject fruits which are fermenting, and that there may be an olfactory basis for such rejection. Although Schaller often stalked mountain gorillas and was upwind of the animals, they rarely showed signs of being aware of him. On one occasion when he had been sweating profusely the animals milled around as if aware of the smell but unable to locate its source. As discussed in Chapter 3, gorillas, particularly the silverbacked males, have a strong body odour which emanates from the armpits where there are large numbers of apocrine glands. This odour is easily identified by human observers and, presumably, also by the gorillas themselves. Hess has noted frequent olfactory inspection of the female's armpits and vagina by the male in association with mating activities, a subject I shall return to in Chapter 7.

The gorilla's sense of hearing is probably quite similar to that of man, or so field observations would suggest. It is quite possible to track gorillas closely without them being aware of one's presence. This is particularly the case if a group is feeding, for the noise of snapping vegetation covers the sound of approaching footsteps. Some of the prosimian primates can hear ultrasonic sounds, but in man, and probably in the gorilla as well, hearing is confined to much lower frequencies. Sound spectrograms of the gorilla's vocalizations also indicate that they are of a low-frequency type.

Of the senses of touch and taste, very little is known. As we shall see in Chapter 6, gorillas feed on the stems, leaves and roots of many plants which we would consider unpalatable. Stinging nettles make no impact on the thick skin of gorillas and they will sit in the midst of such plants, swallowing them with no signs of discomfort. However, Sabater Pi has observed that western lowland gorillas in Rio Muni include a greater proportion of fruits in their diet than mountain gorillas do. He tasted seventy-five per cent of the ninety-two vegetable foods eaten by the gorillas and found that forty per cent of them had a particularly sweet or acid taste. It is interesting that gorillas in captivity learn to accept a whole range of fruits and vegetables in lieu of their normal diet. Their sense of taste may actually be well developed and may be important in discriminating food items in the wild, particularly since some plants are poisonous.

Intelligence

Measuring the intellectual abilities of apes and making cross-species comparisons on the basis of laboratory experiments is a difficult task. Before we consider the results of such studies on gorillas, chimpanzees and orang-utans, it will be as well to make some comparisons of their behaviour in the wild. This will provide a backcloth against which to judge their performances in the artificial testing situations imposed upon them in captivity.

Field studies have shown that, of the three great apes, it is the chimpanzee which shows the most striking examples of tool-use and fabrication. In Tanzania, Goodall observed chimpanzees stripping the leaves from twigs and using them to 'fish' for termites. The modified twig was thrust carefully into a passage in a termite mound, whereupon the insects bit into it and were still hanging on when the twig was withdrawn. For three or four months of each year, the chimpanzees were able to obtain large quantities of termites (*Macrotermes bellicosus*) in this way. The modification of twigs and their manipulation by use of a precision grip involving thumb and index finger is a remarkable example of tool fabrication and use in obtaining food.

Learning by example and practice are crucial elements in the development of this behaviour. Young chimpanzees watch their mothers termite fishing and when seven or eight months old they begin to eat termites occasionally. At eight months an infant may show 'mopping' behaviour as do adults, that is, putting the back of one hand down amongst a mass of termites and allowing them to become entangled in the hair prior to eating them. A one or two-year old infant picks twigs and strips leaves from them and two eighteen-month old chimpanzees were observed jabbing such crude tools into a termite mound. From two years onwards youngsters show more and more elements of the adult feeding pattern but not until five or six years of age do they exhibit full mastery of this feeding method.

This example serves to show the importance of early experience in the development of tool-use in chimpanzees. In laboratory studies by Menzel, Davenport and Rogers in which ten wild-born chimpanzees were compared with eight captive born individuals which had been raised in isolation for the first two years, the wild-born apes far exceeded the isolates in their ability to learn problems involving the use of sticks to obtain food rewards. Some of the results of these studies are shown in Fig. 33.

Termite-fishing has been observed also in an artificial group of chimpanzees in a forest in Gambia. These animals were orphans and came originally from Portuguese Guinea. They also showed the ability to use sticks as levers and to make 'sponges' by chewing up leaves and using them to soak up water from holes in trees. The chimpanzees of Gombe Stream Reserve in Tanzania also exhibit such behaviour and may use leaves to wipe away faeces or blood from their bodies.

Large sticks are also used by chimpanzees to break into termite nests. Goodall observed animals at the Gombe Stream Reserve using sticks for this purpose. Suzuki, working in Tanzania in the Kasakati Basin, observed sticks forty-five centimetres long left in termite mounds by chimps, and Jones and Sabater Pi found 174 such tools, stripped of leaves and twigs, which had been used by chimps in Rio

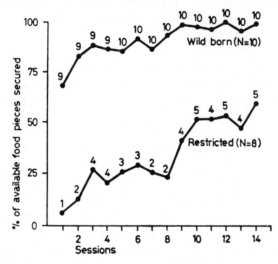

Figure 33. Comparative learning skills of wild-reared and isolation-reared chimpanzees. The problem involved use of a stick to obtain food rewards placed outside the cage. The wild-reared chimpanzees were superior to the isolation-reared animals but the latter improved during the course of the fourteen testing sessions. After Menzel, Davenport and Rogers (1970).

Muni. Another example of the use of sticks is provided by Merfield who many years ago observed chimps in Cameroun using tools to obtain honey from an underground bees' nest. These examples of twig and stick use by chimp populations in various parts of Africa indicate that some cultural transmission of feeding behaviours may occur so that gradually a learned behaviour spreads throughout different groups. Such cultural transmission is not confined, of course, to chimpanzees. One is reminded of other examples such as the habit of washing sweet potatoes which was first learned by young Japanese macaques and then generally spread to other members of their group.

There are examples where particular chimpanzee populations do not show behaviours reported for groups in other areas. Thus Korlandt was unable to find evidence of stick use for feeding by chimpanzees in the Congo, and Reynolds and Reynolds found no signs that chimps fed on termites in the Budongo Forest of Uganda in the same way as Goodall had described for this species in Tanzania. The Gombe Stream chimpanzees do not use leafy twigs as flywhisks, as was reported by Sugiyama in Uganda. Nor do the Gombe Stream chimpanzees use sticks or rocks to break open hard-shelled fruits. Such behaviour has been reported, however, by Struhsaker and Hunkeler, for chimpanzees in the Ivory Coast of West Africa. Chimpanzees visited particular 'nut-smashing places' where they placed the fruits of *Panda oleosa* or *Coula edulis* in depressions in the exposed roots of trees and employed sticks or rocks to break them open.

One further example of the use of sticks by wild chimpanzees comes from the work of Adrian Kortlandt at the Beni Reserve, Northern Kivu Province of the Congo, now Zaire. Kortlandt tested the response of chimpanzees to objects which he placed in their environment. Surprisingly, perhaps, chimpanzees showed little response to a live Gaboon viper or a large dummy spider. When suddenly confronted by a stuffed leopard, however, chimpanzees screamed, rushed about and scratched profusely or defaecated. Some animals charged the dummy, clubbing it with sticks or hurling rocks towards it. Kortlandt interprets such behaviour as supporting his 'dehumanization hypothesis' which proposes that chimpanzee ancestors were once in competition with human ancestors for mastery of the savannah environment but retreated into the forests in view of the superiority displayed by other hominoids. The finding that chimps used sticks as clubs in this way is remarkable but, as Albrecht and Dunnett have pointed out, it is very odd that they never attack real leopards but rather seem to ignore them.

In captivity, chimpanzees have been observed using tools in various contexts. Menzel, working at Delta Primate Centre, observed that young chimpanzees learn to stand loose branches and other elongated objects upright and climb up them. Eventually the animals learned to lean a branch up against a fence 5.5 metres high, and so create a ladder which allowed them to escape from their compound. This behaviour was apparently spontaneous, for no deliberate training was involved, but since the chimpanzees had observed humans using ladders it is possible that they learned partly by observation.

McGrew and Tutin, also working at the Delta Centre observed seven chimpanzees between seven and eleven years old but which had been captured as infants in the wild. One female, named 'Belle', showed an inclination to groom the mouths of other group members, particularly 'Bandit', a young male who was in the process of shedding his milk teeth. Belle usually employed her hands for this operation but on four occasions she used a twig and was once observed modifying the twig to make a more efficient tool.

Gorillas and orang-utans provide a striking contrast to the chimpanzee for neither of them has been observed using tools in the wild. Gorillas sometimes pull-up or shake vegetation during the chest-beating display and orang-utans sometimes snap branches and drop them towards people on the ground, but they apparently do not use sticks or other implements for feeding purposes. A great deal of field data have been collected on gorillas and orang-utans so that it seems likely that neither species uses tools in the wild. Schaller commented on the lack of curiosity shown by mountain gorillas towards novel objects he left in their way such as a rucksack and, in a film of mountain gorillas at Kahuzi, a hat was left in the way of a group as it foraged but only a few of the younger animals showed any interest in it. Gorillas in captivity have not been observed spontaneously making or using tools. Orang-utans by contrast are highly manipulative apes and are well known for their patience and persistence in dismantling objects, including their cages. Orang-utans and chimpanzees are frequently more successful than the gorilla at solving problems that involve manipulating objects, a point to which I shall return in a moment.

It is perhaps as well to remind ourselves also that tool use is not confined to higher primates. Thus one of Darwin's finches *Cactospiza pallida* uses cactus spines and twigs to dislodge insects from crevices in decaying trees. The Egyptian vulture *Neophron pernopterus* smashes open ostrich eggs by hurling rocks at them with its beak, and the sea otter *Enhydra lutris* carries a stone on its chest as it floats on its back and uses this as an anvil against which to smash shellfish. Alcock, in a recent review, makes the interesting point that woodpecker finches, sea otters and man have all developed tool use after invading niches which are not characteristic of their phylogenetic groups. The woodpecker finch is one of the most recently evolved of Darwin's finches, *Enhydra lutris* is the only marine otter and man himself is a recently evolved primate whose ancestors possibly developed tool use in response to selective pressures imposed by competition with already well-established savannah carnivore species. Did similar selection pressures operate to mould the behaviour of chimpanzees, but fail to do so in the case of gorillas and orang-utans since they remained in their forest niches? We cannot know, but if Kortlandt's hypothesis is to be believed, then this may have been the case.

Chimpanzees, unlike the gorilla and orang-utan, kill and eat other mammals. Between 1960 and 1970 the fifty chimpanzees inhabiting the Kakombe Valley at the Gombe Stream Reserve were observed to kill ninety-five animals and another thirty-seven escaped. Fifty-six of the ninety-five prey were identified and these comprised fourteen colobus monkeys, twenty-one baboons, one blue monkey, ten young bush pigs and ten young bush buck. Most of the animals that chimpanzees attacked weighed less than five kilogrammes.

Geza Teleki made an intensive study of predatory behaviour in chimpanzees and observed thirty cases of such behaviour, twelve of which were successful. Only adults hunt, and it is nearly always the males who do so. A male may hunt alone or as many as five may apparently work silently together to surround the prey. The hunters share the food with other group members except for the brain which seems to be a prized item since it is not shared. Chimpanzees may 'request' food from the animal which has possession of the prey by looking into its eyes, reaching out an upturned hand and uttering whimpering noises. Teleki observed 395 instances of requesting behaviour of which 114 were successful.

Baboons also kill and eat mammals such as young bush buck but gorillas and orang-utans have never been observed to eat meat in the wild. The hunting behaviour of chimpanzees is another intriguing example of their intellectual capacities but it is as well to keep it in perspective. Meat forms only a small part of their diet and their co-operative behaviour is less impressive than that of social carnivores such as the lion or hyena.

The abilities of captive apes

We have considered the behaviour of wild apes, and bearing this in mind can now look realistically at the more extensive literature pertaining to intelligence in captive chimpanzees, orang-utans and gorillas. The testing situations devised by psychologists of course make perfect sense to us, but we should not imagine that they are the ideal way of assessing learning and intelligence in apes. Tests which involve the mastery of tool use or a recognition of geometric shapes are simple enough for man to comprehend; gorillas and orang-utans, however, do not normally use tools, and the forest is not composed of stereotyped objects such as triangles and squares. The conceptual world of great apes must be rather different from our own and it is optimistic to expect them to master mechanisms which it took man thousands of years to invent. It is staggering then, that chimpanzees can in fact learn to operate padlocks, type requests on a computer console or communicate with man by sign language.

Classic research on great ape learning was done by Wolfgang

1 Gorilla battling with a leopard; an imaginative illustration from Brehm's *Life of Animals* (1895).

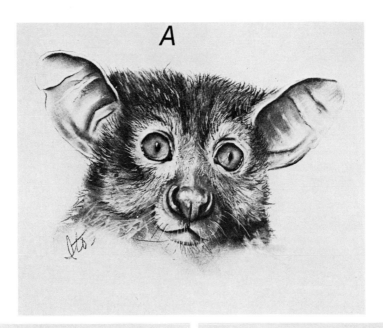

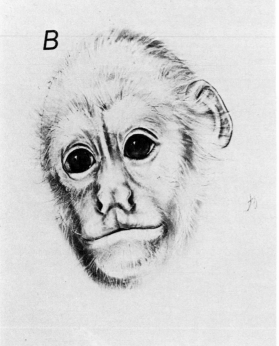

2 Portraits of typical prosimian and anthropoid primates.
A Greater bushbaby (*Galago crassicaudatus argentatus*). Note the large eyes and ears and the moist rhinarium of this nocturnal prosimian; **B** Capuchin monkey (*Cebus*) a platyrrhine primate with widely spaced, laterally directed nostrils; **C** A pigtail macaque (*Macaca nemestrina*) a catarrhine monkey with nostrils situated close together and pointing downwards.

3 A female gibbon (*Hylobates*) with her infant.

4 An adult male Bornean orang-utan (*Pongo p. pygmaeus*) climbing quadrumanously. Inset: detail of the adult male's head to show the cheek flanges and laryngeal sac.

5 ABOVE Chimpanzees. **A** Common
Chimpanzee (*Pan troglodytes*); adult male
(left) female and infant.

LEFT **B** Pigmy chimpanzee (*Pan
paniscus*) female.

6 ABOVE An adult male western lowland gorilla (*Gorilla g. gorilla*) in a knuckle-walking stance. This is an old male (over thirty years) and white hairs have spread from his saddle on to the flanks, buttocks and thighs.

7 LEFT The immunological properties of the blood of man and the apes compared by immunoelectrophoresis. For explanation see text.

8 BELOW 'Congo', a young female mountain gorilla stacking boxes to reach a suspended food reward.

9 Tropical rain forest habitat of *Gorilla g. gorilla* in Rio Muni. Primary forest is visible in the background and secondary forest including *Musanga* trees (M) and *Alframomum* (A) occupies the foreground.

10 Mountain rain forest (*Hagenia* woodland) encircling a grassy meadow at Kabara in the Virunga volcanoes.

11 ABOVE LEFT Bamboo forest.

12 ABOVE RIGHT Giant Senecios, overgrown with mosses and lichens at high altitude in the Virunga volcanoes.

13 BELOW Tree heath covered by lichens at high altitude in the Virunga volcanoes.

14 Mountain gorillas in a *Hagenia* tree. An adult female is visible in the foreground. Thick pads of moss cover the branches of blade-shaped *Polypodium* ferns hang underneath. Gorillas sometimes feed on these ferns.

15 ABOVE Four views of a mountain gorilla carefully removing and eating the bark of a vine (*Urera hypsolendron*).

16 BELOW Night nest built by a mountain gorilla. The nest is on the ground and contains segments of dung.

18 ABOVE Part of a group of mountain gorillas. The silverback in the foreground displays with a 'strutting walk' in response to the observer.

17 OPPOSITE A group of mountain gorillas on the move, with two adult males in the lead.

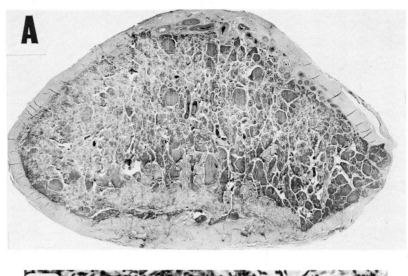

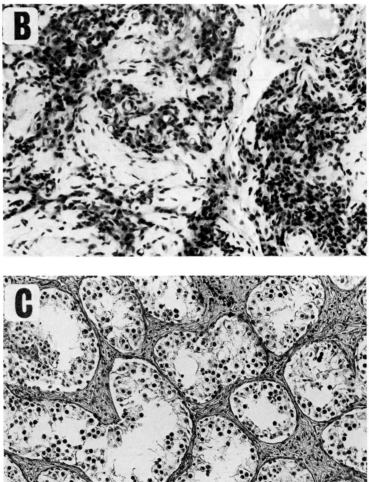

19 Testicular atrophy in captive gorillas. **A** L.S. of testis of 'Guy' (x5) to show the diffuse nodular appearance of the tissues; **B** Testis of 'Guy' (x160). The seminiferous tubules are degenerate and there is an extensive stroma of interstitial cells; **C** Testis of 'Oban' (x160). Some sloughing and degeneration of the seminiferous epithelium is apparent.

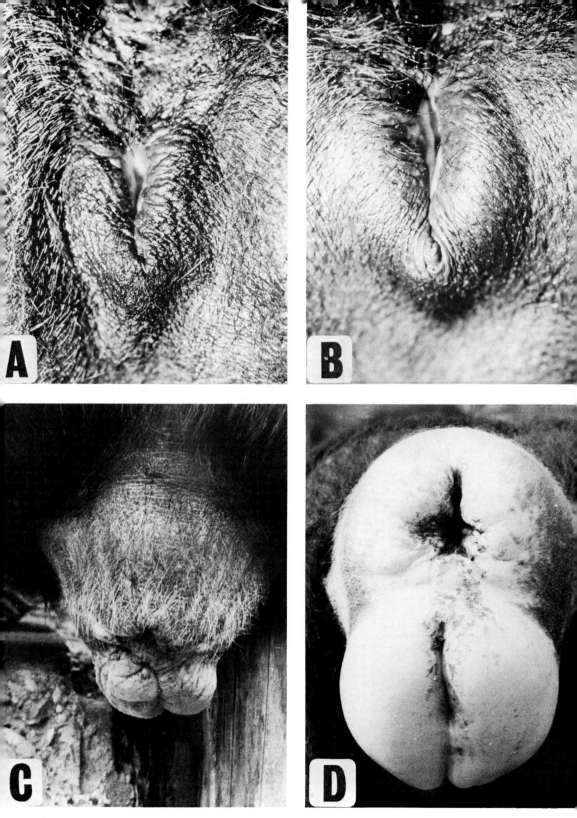

20 External genitalia of the non-pregnant female gorilla and chimpanzee.
A Female gorilla: labia show minimal swelling; **B** Female gorilla: maximal labial
swelling at mid-cycle; **C** Chimpanzee: sexual skin detumescent; **D** Chimpanzee:
sexual skin fully tumescent at mid-cycle.

21 'Salome', an infant female western gorilla at London Zoo, after removal from her mother at six weeks of age.

22 'N'Pongo', a female western gorilla with her infant. Her first three infants had to be removed for hand-rearing.

23 Hand grasp reflex of a newborn female gorilla.

24 The Virunga volcanoes – one of the last refuges of the mountain gorilla. In the foreground is Mt Visoke and behind it, Mt Karasimbi.

Köhler and Robert Yerkes during the 1920s and 1930s. It is impossible to pass over these pioneering studies or to dicuss the intelligence of the gorilla in isolation since most of the relevant experimentation was conducted using chimpanzees.

During the First World War Wolfgang Köhler was in charge of a primate research unit on the island of Tenerife in the Canaries. In 1914 he conducted experiments on the intelligence of nine chimpanzees in his care. One of the animals was an adult female but the other eight were youngsters aged between four and seven years and all had been born in the wild, in Cameroun. Köhler investigated the intelligence of these apes in a series of problems, the solution of which would enable them to reach a food incentive. The first set of problems involved the use of 'roundabout methods' to reach the food reward. An example is quoted below.

The objective hangs in a basket from the wire roof and cannot be reached from the ground; the basket contains several heavy stones also, so that one push of the string and basket will make the whole swing for some little time; the swing is so arranged that the longest sideways movement of the basket makes it nearly reach a scaffolding. Thus the roundabout way is easily recognisable and available, but only for a few moments. As soon as the basket is swinging, Chica, Grande and Tercera are let in upon the scene. Grande leaps for the basket from the ground and misses it. Chica who in the meantime, has quietly surveyed the situation, suddenly runs towards the scaffolding, waits with outstretched arms for the basket and catches it. The experiment lasted about one minute.

This excerpt demonstrates one of Köhler's major findings: that a chimpanzee could in some cases show a flash of 'insight' which allowed it to proceed smoothly to the solution of a problem without fumbling upon it by trial and error. In the case of the problem above, all the chimpanzees eventually solved it, including 'Grande', whom Köhler found to be 'always slower than the others'. Köhler was well aware that there were large individual differences in the abilities of his nine subjects and that it was difficult to provide a simple explanation for these. Thus, one animal might be innately more gifted than another, or, on the other hand, early experience might have been the deciding factor, for he knew nothing of these chimpanzees' earliest days in the forests of Cameroun.

Köhler also tested chimpanzees in problems where it was necessary to manipulate strings in order to obtain a food reward. The simplest form of this task involved tethering a chimpanzee to a tree so that his range of movement was limited, and then placing some food, usually bananas, just out of his reach. A string was attached to the bananas, however, and the chimp could easily reach the free end of this and draw in the food. Köhler's animals swiftly mastered this problem, and he then set more difficult forms of the task in which apes were given a choice of several strings which criss-crossed one another but only one of which led to the food. The chimpanzees found this problem much more difficult. Thus Sultan, who excelled in many tests, acted hastily and often chose the wrong string. Köhler commented, however, that 'his mistakes can scarcely be fortuitous, in the course of five experiments he gave the first pull four times to the string, which appeared to reach the objective by the shortest distance.' It is interesting to consider the perceptual processes of the chimpanzee in this situation, for Köhler noted that the ape 'always pulls the string if it visibly touches the objective. It appears doubtful whether the concept of "connection" in our practical human sense signified more for the chimpanzee than visual contact in a higher or lower degree.'

This method of testing has been modified and used extensively since Köhler's time. Finch, in 1941, tested eight chimpanzees on eleven variants of the problem and concluded that their performances were superior to those of rhesus monkeys. In 1953, Reisen and his co-workers found that young gorillas achieved results similar to those of Finch's chimpanzees and this result was later confirmed by Fischer and Kitchener who also reported that orang-utans performed similarly to the other two great apes.

Köhler also developed two other techniques which have been used extensively by later workers; testing the ability of chimpanzees to use sticks in obtaining food rewards, and to stack boxes and so obtain an objective when it hung above them and out of reach. At first Köhler gave chimpanzees the opportunity to use a single stick to rake in bananas placed near their cages. The following account concerns Nueva, a female, tested three days after her arrival at the field station.

A little stick is introduced into her cage; she scrapes the ground with it, pushes the banana skins together into a heap and then carelessly drops the stick at a distance of three quarters of a metre from the bars. Ten minutes later, fruit is placed outside her cage beyond her reach. She grasps at it, vainly of course, and then begins the characteristic complaint of the chimpanzee; she thrusts both lips – especially the lower – forward, for a couple of inches, gazes imploringly at the observer, utters whimpering sounds, and finally flings herself on her back – a gesture most eloquent of despair, which may be observed on other occasions as well. Thus between lamentations and entreaties some time passes until – about seven minutes after the fruit has been exhibited to her – she suddenly casts a look at the stick, ceases her moaning, seizes the stick, stretches it out of the cage, and succeeds, though somewhat clumsily, in drawing the bananas within arms' length.

Since Nueva was born in the wild, she had some previous experience in manipulating sticks, if not actually using them as tools. Again we see the operation of what Köhler calls 'insight' in a rapid solution of the problem and indeed when the task was set again one hour later she completed it much more rapidly and skilfully, indicating an ability to profit from experience and practice. Not all the chimpanzees were so successful, however; a young male called Koko was clumsy in his use of the stick and pushed the fruit away instead of drawing it to him, or used his foot to grip the stick instead of his hand. Köhler's description of Nueva's emotional behaviour during testing is also interesting; chimpanzees are prone to exhibit such behaviour even to the extent of throwing tantrums, whereas the gorilla is much more methodical and placid in its responses.

Köhler was able to set more complex forms of the stick problem to chimpanzees once they had mastered the use of a single stick. Thus animals learned to use two sticks which fitted together at the ends and so reach bananas too far away to be secured with one stick alone. One male named Sultan learned to fit three such sticks together.

Köhler noted that in learning the solutions to these tests an important factor was where the implement was placed relative to the food objective. If the stick was placed at the back of the animal's cage and not in view of the fruit, the chimpanzee often failed to use it to secure the reward. Some animals, however, had clearly mastered the principles involved in solving the problem; for instance, Nueva, who used rags, straws, or her tin drinking bowl to extend her reach and so draw in bananas when a stick was not available.

One of Köhler's most famous testing methods, and one later applied to orang-utans and gorillas, concerned the use of boxes as a means of obtaining a food objective hanging high above the ground and out of reach. In the simplest stage of this problem the bananas were arranged so that only one box was required to reach them. Gradually, however, Köhler extended the problem until three boxes were needed, stacked one on top of another.

The objective hangs still higher up; Sultan has fasted all the afternoon and, therefore, goes to his task with zeal. He lays the heavy box flat under the objective, puts the second one upright upon it, and, standing on the top, tries to seize the objective. As he does not reach it, he looks down and round about, and his glance is caught by the third box, which may have seemed useless to him at first, because of its smallness. He climbs down very carefully, seizes the box, climbs up with it and completes the construction.

Some chimpanzees were not successful in this task, however, for Köhler notes that 'Konsul never built, Tercera and Tschego got no further than some feeble attempts.' Some of the stacks of boxes constructed were obviously unsound, yet animals attempted to use them. On one occasion he observed a female named Rana attempting to obtain a suspended banana by holding two sticks end-on-end and then trying to climb up them. He considered that chimpanzees appreciated only the optical aspects of such situations but failed to realize the 'technically physical point of view'.

In 1949 Harry Harlow published a now classic paper entitled 'The formation of learning sets'. He demonstrated that in a series of similar problems, learning occurred mainly on the first trial and that the experience gained by rhesus monkeys enabled them to perform more successfully on later problems. Köhler did not emphasize the role played by experience in his experiments, but it is certain that he must have realized its relevance, for in the patterned string, stick and box stacking experiments he tested the apes on problems of graded intensity starting with the simplest case and gradually making matters more difficult.

Fischer studied learning-set formation in two gorillas aged seventeen and twenty-one months, testing them on 232 pattern-discrimination problems. She found that the apes did slightly better than rhesus monkeys at a similar developmental stage. In 1967 Rumbaugh and McCormack published a comparative study which indicated that the learning-set skills of gorillas are similar to those of chimpanzees and orang-utans but greatly exceed those of the gibbon or squirrel monkey.

Whilst Köhler was writing up his studies on the chimpanzee, he received a monograph from Robert Yerkes in which this researcher formed similar conclusions about the learning capacities of great apes. Yerkes had studied an orang-utan and attributed to the animal the ability to solve problems by 'ideation', the same process which Köhler had called 'insight'.

Orang-utans are highly manipulative and curious creatures and Hornaday in his *Minds and Manners of Wild Animals* recorded that his orangs were trained to 'ride a tricycle or bicycle', or to 'drive in nails with a hammer' and even to 'put on a pair of trousers, adjust the suspenders, put on a sweater or coat and a cap'. Such tricks are distasteful in that they cast apes in the role of quasi-humans and in any case they tell us little regarding the true nature of these animals.

Yerkes carried out his investigations using a male orang-utan named Julius. He set Julius a box and suspended a banana problem similar to that employed by Köhler but the situation was a difficult one involving two boxes without any previous opportunity to use a single box to reach the objective. Julius did not stack the boxes, but did attempt to reach the food by climbing the walls of the cage and even by leading Yerkes under the suspended food and attempting to climb up the experimenter! Yerkes recorded 'I was amazed by Julius' behaviour this morning, for it was far more deliberate and apparently reflective as well as more persistently directed towards the goal than I had anticipated.' Eventually, when Julius was shown how to stack three boxes and allowed to obtain the food, he learned the solution, and from then on performed the task faultlessly.

In another test, Yerkes compared the performance of Julius with that of monkeys, crows, rats and pigs in a multiple choice problem. This consisted of a row of nine boxes with each having an entrance and exit door, one of which contained the food reward. In the first instance the reward was concealed in the first box on the left from the starting point of the test. The two monkeys learned the problem (Fig. 34) making a high percentage of errors at first, and arriving at the solution by trial and error. Julius, however, took 290 trials to master the apparently simple problem and so fell far behind even the crows and pigs tested by Yerkes. Julius, however, did not work by trial and error, but adopted a definite method, moving always to the nearest box from the starting point but not necessarily the one to the left of him. After 230 trials however, Yerkes reset the problem so that the nearest box was never the right one and Julius was consistently wrong. As can be seen in Fig. 34 Julius then mastered the problem so that his errors declined suddenly to nil and Yerkes was convinced that 'the orang-utan solved this problem ideationally'.

One important point which emerged from this experiment is that the number of trials taken to solve such a problem is less important than the method of solution. Thus, although the two monkeys in Fig. 34 reached the answer much more rapidly than Julius, the ape clearly exceeded them in intelligence.

Yerkes considered it essential in his experiments with great apes that he should get to know the behavioural patterns of individual subjects as completely as possible. Nowhere is this approach better exemplified than in his dealings with Congo, a young female mountain gorilla which Ben Burbridge had brought back from the Virunga volcanoes. Yerkes studied this animal during three consecutive summers, first at Burbridge's home and finally at Ringling Brothers' Circus. Indeed if Congo had not died, Yerkes had intended to continue his studies.

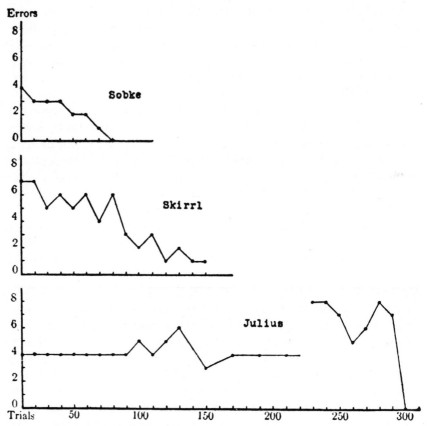

Figure 34. Comparison of the learning skills of an orang-utan (Julius) and of two monkeys. For explanation see text. After Yerkes and Yerkes (1929).

In 1926 when the research began, Congo was about four or five years old and weighed sixty-five pounds, but by 1928 her weight was 160 pounds and she was reaching puberty. Many variables may have affected Congo's performance in behavioural tests, including her increasing physical and mental development and also her memory of testing in previous years. The change of environment may also have affected her behaviour during the final year of the study, when she was purchased by a circus. Bearing in mind Köhler's finding that there was great individual variability in intelligence among chimpanzees, it is important to remember that Yerkes' results apply to only a single gorilla.

I preface my description of Yerkes' experiments in this way because Congo's performance in many tests was very poor when compared to chimpanzees and orang-utans, and because she sometimes failed to solve problems in which she had succeeded during the previous year.

She was clumsy in her attempts to use a stick to rake in a food reward, but she did work consistently and calmly without the emotional outbursts typical of chimpanzees. Often she swept the stick to the left irrespective of the position of the objective, and was never able to manipulate it smoothly. Likewise Congo's performance in box-stacking problems was very poor though she did eventually achieve success (see Plate 8). In another problem, Congo was required to push a pole through a long hollow container, open at both ends, and so push out some food concealed inside it. Congo did succeed on this problem but in the final year of testing she failed to do so, and in any case Yerkes considered her performance to be much inferior to that of the chimpanzee or orang-utan. In what seems a very simple problem by human standards, Yerkes wound Congo's tethering chain round and round a tree trunk so that her sphere of movement was restricted. In order to reach some food placed on the ground nearby, Congo had to retrace the path of the chain and so untangle it, but she seemed to be baffled by the situation.

In many of these problems it may have been Congo's lack of manipulative ability which hampered her, rather than a lack of actual intelligence. Thus, she was able to learn to remove an open padlock from a hasp and so obtain food from a box, but 'the operation was not deft and skilful as in the case of chimpanzee or man, but instead a crude, clumsy fumbling'.

In a problem involving colour discrimination and delayed response, however, Congo fared rather better than in tests based mainly on manipulative ability. In this experiment she was able to recognize and remember which coloured container out of five on a turntable had food inside it. Chimpanzees also performed well on delayed response problems, for instance one subject tested by Yerkes had no difficulty remembering which one of four boxes had been baited with food four hours previously.

The reason why gorillas do not perform as well as chimpanzees or orang-utans in problems involving tool use may be because their manipulative abilities are inferior and their inclination to manipulate objects is less than in the other great apes. This may correlate in some way with ecological factors; gorillas are mainly terrestrial creatures feeding upon plants which are usually easily obtainable. This is not to ignore the fact that gorillas can climb trees remarkably well, that their ancestors were arboreal and that in the gathering of their food the hands are used a great deal and quite dexterously. Chimpanzees and orang-utans, however, feed habitually in the trees and in order to reach fruit must employ a good deal of care and intelligence. They are heavy animals and many items of food are not easily reached. Probably orang-utans and chimpanzees employ 'roundabout methods' as an everyday part of their foraging behaviour, and, in the case of the chimpanzee, one remembers that this species quite often uses simple tools to obtain food.

Due to lack of appreciation of these factors, a mistaken view developed in the past that chimpanzees were the most intelligent of the three great apes, with orang-utans in second place and the gorilla a rather poor third. Actually, it is possible to bias the testing situation so that a reversal of the rank order occurs, as has been shown by Rumbaugh, Gill and Wright. In these experiments, chimpanzees, gorillas and orang-utans were trained to criterion (that is, no more than three errors out of ten choices) on a set of two-choice object quality discrimination problems. The pairs of objects to be discriminated were held in glass-fronted bins. When criterion was achieved a half-inch wire mesh was placed over the glass front of each bin so that the apes had to look through this at the test objects. Under these conditions the performance of the orang-utans suffered most, with a thirty per cent drop on the test trials of the first ten problems. Chimpanzees showed a twenty per cent drop and gorillas only a ten per cent drop. Further, gorillas reached criterion with the first eleven problems but orang-utans needed twenty-two problems and chimpanzees fourteen (Fig. 35). It has been suggested that the reason

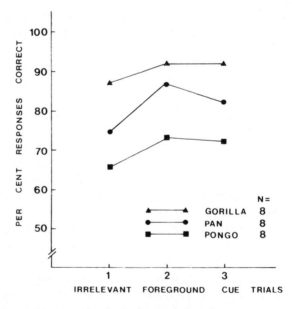

Figure 35. Effects of 'irrelevant foreground cues' upon the learning performances of gorillas, chimpanzees and orang-utans. For explanation see text. Redrawn from Rumbaugh (1974).

orang-utans and chimpanzees pay more attention than gorillas to 'irrelevant foreground cues' is that they are arboreal creatures and need to avoid twigs and leaves which might damage their eyes. This is an interesting suggestion, but a difficult one to evaluate; gorillas are certainly not as arboreal as the other great apes, but they do, none the less, live in dense, and often prickly, vegetation.

A number of studies indicate that the intelligence of the gorilla is at least equal to that of the chimpanzee and orang-utan. In 1942 Benchley described some experiments which Harold Bingham had carried out, using two gorillas at the Yerkes Primate Laboratories. The testing apparatus was a long box closed at one end and with a number of longitudinal slots in its sides. Food was placed at the closed end of the box and in order to obtain it, the apes would have to move it along to the open end by poking their fingers through the slots. Yerkes had previously set this problem to Congo, and her performance had been very poor. The gorillas studied by Bingham, however, did much better. The male, Mbongo, first solved the problem, working quietly and methodically to obtain the fruit but allowing the female, Ngagi, to eat it! The roles were later reversed, however, and Ngagi obtained the fruit but allowed the male to consume it. This is a remarkable example of altruistic and co-operative behaviour. It brings to mind some work done by Crawford in which two chimpanzees learned to co-operate with one another to pull on a rope and so draw a box containing food towards their cage. The box was too heavy for either chimpanzee to manage alone.

Bingham found that the gorillas' approach to the slot-box problem was superior to that of the chimpanzees and orang-utans he tested. The gorillas were more patient and persistent than the chimpanzees and did not resort to attempts to smash up the apparatus as the orang-utans did.

Rumbaugh and McCormack have compared the performances of gorillas, chimpanzees and orang-utans in oddity concept problems and have also tested gibbons and rhesus monkeys. In such problems the animal must choose the odd-man-out of a group of stimulus objects. Thus, if size is the cue involved, all the objects presented may be different shapes, but all of them except one will be the same size. The great apes performed at sixty-five to ninety per cent correct levels 'there being great overlap between the three genera'. Macaques also did well on these problems but of the gibbons only one showed evidence of having mastered the oddity concept and then it was only forty per cent correct on the final tests.

Gibbons are something of a mystery, for their learning skills in many types of tests are poorer than those of monkeys. It may be, for reasons not understood, that the testing situations employed are not suitable. Rumbaugh notes, for instance, that young gibbons can be very responsive and alert. Also, he observed an adult male who, when his chain got twisted round the cage supports, retraced his path delicately with the chain in one hand so as to avoid getting it tangled. Yet when Yerkes had set such a 'wound chain problem' to a young female gorilla, she was apparently baffled.

There is no doubt that all the great apes possess learning capacities much superior to those of monkeys. *Cebus* monkeys, as Klüver demonstrated in the 1930s, are exceptionally intelligent and perform well on problems involving tool use; but both behaviourally and neurologically the apes are closer to man. Yerkes considered that the gorilla was superior to the orang-utan and chimpanzee in its ability to show ideational learning, but that chimpanzees exceeded orang-utans and gorillas in trial-and-error learning. Chimpanzees were the most curious of the apes, but gorillas, it seemed, often lacked the motivation to investigate novel objects and take part in testing procedures. Actually, Rumbaugh has pointed out that when factors such as age, experience, individual variability and biased testing methods have been taken into account, the three great apes do equally well. His work has involved the use of a measure called the 'transfer index' which he considers the most accurate way to make phylogenetic comparisons of the abilities, rather than just the performances of primates in learning tests.

In the transfer index (TI) procedure, animals were first trained on problems involving visual dicrimination of a pair of stimuli, one of which was rewarded as the correct choice. Once criterion had been reached at either the sixty-seven per cent or eighty-four per cent level, as long as this occurred in no less than eleven and no more than sixty trials, then the problems were reversed. The previously correct stimulus in each pair was now incorrect and the percentage of correct choices ($R\%$) was computed in these post-reversal tests. The TI was then calculated by dividing $R\%$ by the pre-reversal criterion, i.e. sixty-seven per cent or eighty-four per cent. Fig. 38 shows transfer indices plotted for a range of primates, including all the great apes, gibbons, the vervet monkey, the talapoin monkey and the ring-tailed lemur.

As can be seen, the TI capabilities of gorillas at both sixty-seven and eighty-four per cent pre-reversal criteria are very similar to those of orang-utans and chimpanzees. The great apes are superior to gibbons, monkeys and lemurs, and a gradation in performance occurs which mirrors the phylogenetic scale; the great apes are at the top and the lemur is at the bottom.

As can be seen in Fig. 36, whereas gorillas and other great apes benefited from pre-reversal training and rapidly learned that the problems had been altered, the talapoin showed negative transfer of learning. That is to say, the more talapoins had learned, the worse they did when the problems were switched around.

A number of other studies indicate the extent to which the great apes exceed monkeys in their intellectual performance, but, unfortunately, most experimentation has been attempted with chimpanzees

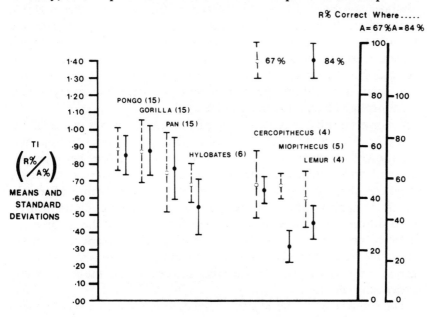

Figure 36. A comparison of the transfer indices of a range of primates, including gorillas, orang-utans and chimpanzees. For explanation see text. Redrawn and modified from Rumbaugh (1974).

or orang-utans rather than with the gorilla. Thus Davenport and Rogers have shown that chimpanzees and orang-utans possess the ability to use visual cues to solve tactile discrimination problems. The animal was allowed to see an object and to reach through an opening in a partition and feel two other objects, one of which matched the visual stimulus. Chimpanzees and orang-utans were able to learn to choose the matching object by touch; a problem which monkeys did not solve.

Chimpanzees have also been shown to be capable of learning the rudiments of language; a capacity which was hitherto believed to be unique to humans. Hayes and Hayes attempted to teach a young female chimpanzee, Viki, spoken language, and the Kelloggs brought up a baby chimpanzee in their home side by side with their own child. Neither animal learned to talk and the most that Viki achieved was to make sounds approximating to 'cup' or 'mamma' in the appropriate contexts.

Beatrice and Allen Gardner adopted a different and much more successful approach to the study of language acquisition in the chimpanzee. Since chimpanzees normally employ a rich repertoire of visual displays for communication, the researchers attempted to teach American Sign Language (ASL) to a female infant chimpanzee called Washoe. Washoe had been captured in the wild and her training began when she was already between eight and fourteen months old. A team of researchers worked with Washoe, using only ASL in her presence, and she was housed in a well-furnished trailer with a large garden. Washoe received every encouragement to develop ASL and after thirty-six months she was using eighty-five formal signs plus two more which she had invented herself.

Washoe's ability to use signs in their correct contexts was tested in a variety of ways. In one test three experimenters worked together; one showed Washoe slides on a screen, a second asked her what she saw and wrote it down, whilst a third worker observed the ape's responses through a one-way mirror, but could not himself see the screen. In such a test ninety-nine slides were shown depicting thirty-three different items, Washoe named fifty-three of them correctly. Some of her errors in tests are interesting. For instance, when she was shown small models of animals, Washoe sometimes made the ASL response for 'baby', perhaps naming the objects on the basis of size rather than identity.

Ten months after the project started, when Washoe was between eighteen and twenty-four months old, she began to make combinations of two signs; human babies also begin to make two word combinations at about this age. Nine months later, Washoe began to employ the pronouns 'you' and 'me' in combination with other signs and during a twenty-six month period she made 245 combinations of three or more signs. Washoe's communication, consisting as it did of phrases such as 'please sweet drink, hurry sweet drink, please gimme sweet drink' etc. may seem banal, but it is remarkable when one considers that the animal had acquired language and this was hitherto believed to be impossible for apes. The Gardners point out that Washoe had no other chimpanzees to interact with and that the experimenters were not fluent users of ASL. They believe that 'under more favourable conditions a chimpanzee could achieve much more'.

A project has also been carried out to study the ability of a young gorilla to learn American Sign Language. The gorilla was born at San Francisco Zoo, taken from its mother, and used in similar experimental procedures to those employed by the Gardners. Since in normal circumstances, gorillas employ gestural communication much less than chimpanzees, one might argue that they would not be so successful in learning ASL. This is not the case, however, for the young gorilla produced results similar to those obtained by studying Washoe.

In man, the capacity for language has often been linked to lateralization of brain function, so it is pertinent to ask whether there is any evidence for such lateralization in the gorilla's brain. There is, in fact, a small amount of behavioural and anatomical evidence. Mountain gorillas observed by Schaller showed a consistent preference to begin their chest-beating displays with the right hand. In eight males, eighty-two per cent of displays began in this way. Groves and Humphrey have shown that in mountain gorillas the left side of the skull is often longer than the right side. This also applies, to a lesser extent, to the eastern lowland gorilla but not to the western subspecies. Groves and Humphrey emphasize that this does not mean that the left cerebral hemisphere is larger than the right one in mountain gorillas. It is more likely that the skull assymetry results from behavioural causes, such as the gorillas habitually chewing their food on the left side.

Besides the work on ASL described above, there have been two other attempts to study language acquisition in the great apes. Both studies were on chimpanzees and both employed communication by manual rather than by vocal means. David Premack began experimenting with a wild-born female chimpanzee when she was six years old and succeeded in training her to communicate by means of plastic symbols. Each magnetized symbol represented a word, and symbols were arranged in vertical rows on a metal slate so as to form phrases and sentences. In the second study, Duane Rumbaugh and his colleagues at Yerkes Primate Centre worked with Lana, a two-and-a-half-year-old chimpanzee. The animal was housed in a seven foot cube with no windows but with a console consisting of seventy-five keys; this provided Lana with a means of communication with the outside world. Each key had a geometric design or lexigram upon it which represented a word and Lana's key-pressing performances were monitored by a computer. Each time Lana composed a sentence correctly, a tone sounded and a reward was given, such as food, toys or the opportunity to look out of a window. In this study, as in the previous ones, it became apparent that the chimpanzee does have the capability to use words, build phrases or simple sentences and use them in their proper contexts. The failure to teach apes vocal language might mean that the larynx is not structured or innervated in a way which is compatible with the production of spoken words but clearly does not mean that great apes lack the brain development to acquire a rudimentary concept of language.

As a final example of the abilities of great apes I shall describe some observations on drawings and paintings done by chimpanzees and gorillas. The observation that great apes will spontaneously indulge in such behaviour is probably very old. Paul Gervais in his *Histoire Naturelle des Mammifères* published in 1854 includes a wood cut of a chimpanzee scribbling on a piece of paper. Köhler also observed how his chimpanzees gradually developed the habit of chewing white clay and smearing it upon objects with their lips. The behaviour was self-rewarding, for the animals seemed to treat it as a game.

Desmond Morris tested the artistic abilities of a number of young chimpanzees; the most outstanding animal was a young male named Congo. Many others have carried out similar studies, including Kohts, Schiller and Hediger and a few general points about their experiments may be made. The scribblings of apes are apparently not totally random. They show a tendency to confine themselves to the sheet of paper supplied and if a central figure such as a square has been placed there beforehand, they tend to concentrate their efforts in this area. If the premarked figure is on the left or right side of the paper, however, the subjects often mark the opposite side of the sheet as if to balance the composition. Morris noted a strong tendency for chimpanzees to paint fan-like patterns of lines. The highest level of composition achieved was in a painting by Congo in which he made a circle and marked inside it. In human infants this stage is the forerunner of pictures of the face.

As with all other forms of testing described in this chapter few gorillas have been studied, but the results seem to be generally similar to those obtained from chimpanzees. Redshaw has observed, however, that when infant gorillas are supplied with drawing materials they show no inclination to use them spontaneously in the way that human babies do. Hediger did some work with Achilla, an adult female western gorilla at Basle Zoo. Unfortunately however, though she produced pictures similar to those of chimpanzees, Achilla's career was cut short, for she swallowed her pencil and surgery was required to remove it from her stomach.

Chapter 6

Behaviour and Ecology

UNTIL recently, very little was known about the life of gorillas in their natural environment and native folklore or the chance observations of travellers and hunters substituted for scientific information. This is not surprising when one considers that gorillas are rare creatures, inhabiting dense forests where great patience is required to observe them.

During the 1930s a number of attempts were made to study both chimpanzees and the gorillas in the wild, notably by Nissen, Pitman and Bingham. The latter published a monograph entitled *Gorillas in a Native Habitat*. These field studies were only partially successful, however, though much information was obtained by examining signs left by the apes such as remains of food items or the nests which they built each night. These pioneering workers usually travelled with large, noisy safaris and were not experienced in field techniques. They often frightened the animals away before much could be learned about behaviour.

The late 1950s saw a resurgence of interest in fieldwork on the African apes, with Kawai and Mizuhara from Japan, Donisthorpe and Osborn from England and the Americans Emlen and Schaller turning their attention to the mountain gorilla, whilst a Spaniard, Jorg Sabater Pi, observed western lowland gorillas in Rio Muni. Schaller's researches in particular provided the first extensive information on wild gorillas, and his book on their ecology and behaviour is an acknowledged classic. Schaller realized that the best way to get close to wild gorillas was to go in search of them unarmed and alone, or with a native tracker. He wore drab clothing in order not to alarm the gorillas and often seated himself in full view of the animals. In this way Schaller found that gorillas gradually accepted him as a harmless addition to their environment. Gorilla groups can, therefore, become habituated to the presence of man, but it has been observed that they tolerate white men more readily than they do the indigenous peoples, probably because it is mainly the latter who hunt for gorillas nowadays.

In recent years field workers have continued to devote most attention to the mountain gorilla. Fossey, Harcourt, Caro, Elliott and Stewart have studied the population of the Virunga volcanoes, whilst Michael Casimir and Alan Goodall have observed the mountain gorillas of Kahuzi. The two lowland subspecies of gorilla are poorly studied. Schaller has provided some information on eastern lowland gorillas, whilst Clyde Jones and Jorg Sabater Pi have published a monograph on the western gorilla. The latter authors observed western gorillas during only forty-six hours of direct contact, however, as compared to the thousands of hours which have been achieved with mountain gorillas. Further, these two scientists had to abandon their work when Rio Muni became independent and foreigners were ejected from the country. The research I shall review in this chapter refers, therefore, mainly to wild mountain gorillas, augmented by some field information on the two lowland subspecies and some behavioural work on captive animals.

The gorilla's habitat

Tropical rain forest has an extensive distribution in Africa, as can be seen in Fig. 37. Schaller has recognized three types of forest which provide a home for gorillas; lowland rain forest, mountain rain forest and bamboo forest (see Plates 9, 10 and 11).

The number of tree species which occur in lowland rain forests is extraordinary. Richards points out that there are seldom fewer than forty species of tree with a trunk diameter of more than four inches in each hectare of a tropical rain forest and sometimes more than a hundred species. Wallace in 1878 wrote:

> If the traveller notices a single species and wishes to find more of it, he may turn his eyes in vain in every direction. Trees of various forms, dimensions, and colours are around him, but he rarely sees any of them repeated.

The trees tend to have tall, straight trunks frequently supported by plank-like buttresses. They vary greatly in height (up to fifty metres) creating strata within the forest. The canopy contains a wealth of epiphytes, such as the lianes which are so typical of tropical rain forest. The lower strata, of great importance to the ground-dwelling gorilla, tend to be sparser. Thus Richards writes of the Nigerian rain forest: 'The shrub stratum is very indefinite ... and in some places the shrub and ground layers are almost wanting. Large stretches of

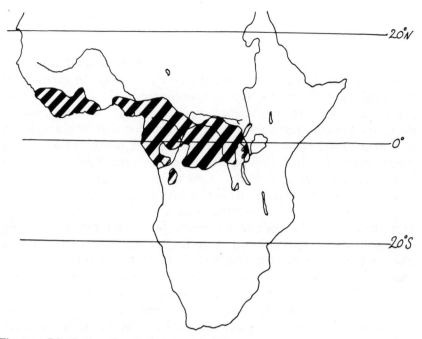

Figure 37. Distribution of tropical rain forest in Africa. Based upon Richards (1972). This map may be compared with Fig. 5 (page 17) which shows the recent distribution range of the gorilla and chimpanzee.

the forest floor may be almost completely bare.' This situation is frequently disturbed, however. Openings in the forest are created by storms, fires, and simply because the great trees die and make gaps as they fall. Traditionally, the most important cause of primary forest destruction has been the pattern of shifting cultivation practised by the human inhabitants. Clearing is usually by the 'slash and burn' method which spares nothing except an occasional large tree. Crops are planted among the unburnt debris. Many factors, including the high oxidation rate of humus in the tropics and leaching and erosion of unprotected soils, contribute to a rapid decline in fertility of the cleared areas. The farming villagers therefore abandon their plots every few years and clear another area. The neglected plots are then subject to the natural regenerative process of 'secondary succession' by which the climax forest may eventually be restored.

Secondary forest (as the regenerating plant community is called) is a habitat much favoured by gorillas, so the stages of secondary succession are worth mentioning. When cultivation ends, the ground is bare except for any spared forest giants and perennial crops such as bananas. After a few weeks the ground is covered by low, dense vegetation comprising herbaceous plants typical of cultivated areas, plants, mostly woody, characteristic of secondary forest but also found in openings of the primary forest, and seedlings of the primary forest trees. During the stages of succession, species from these three groups of plants become dominant in turn. Among the herbaceous plants, Sabater Pi and Jones noted that *Aframomum* (a member of the ginger family) usually becomes dominant as the last of the crop plants (often bananas or manioc) are crowded out. The first woody plant to establish itself above the low vegetation may be a species of *Vernonia*, or a *Trema* species. It may be a matter of chance which species assumes initial dominance, but subsequently the parasol tree (*Musanga cecropoides*) is likely to colonize the regenerating forest. Ross found that by three years it had become dominant in each area he studied and that this dominance lasted for fifteen to twenty years. During this time, the open *Musanga* canopy is about fifteen to twenty metres above the ground and there is probably a second storey of trees and shrubs at four to five metres. Herbaceous ground cover is plentiful, especially in the early stages, when it forms a dense carpet. After fifteen to twenty years, the *Musanga* trees gradually begin to die out. Since *Musanga*-dominated forest is unsuitable for the growth of *Musanga* seedlings, they are replaced by the young trees which had formed the second storey. Initially, these species will include types characteristic of both primary and secondary forest. As regeneration proceeds, the proportion of the former will increase, while that of the latter will decrease, until the primary rain forest is restored.

Regeneration and secondary succession is a huge and fascinating topic. It is important because of the scale on which it is occurring as a consequence of man's activity and, from our point of view, because it creates a forest of a type ideal for the gorilla. It is important to remember, however, that the stages outlined above represent a steady progression towards the re-establishment of the original primary forest, and assume a long interval (more than fifty years) between periods of cultivation. If the interval is short, however, succession may be 'deflected' or 'regressive', leading ultimately to grass savanna. Excessive felling of trees for timber also occurs in many areas and there is insufficient time for the normal succession to occur.

Mountain rain forest, the second major forest type inhabited by gorillas, occurs in more rugged, broken country than its lowland counterpart and at altitudes of from 1,350 to 3,200 metres (see Plate 10). Thus in the Kayonza forest, in southwest Uganda, mountain rain forest occupies the ridge crests whilst the ravine slopes are covered with dense growths of herbs and shrubs. In mountain forests the canopy usually averages about twenty-five metres in height, and there may be a lower tree stratum. The tree crowns are compact and separate. Another striking difference between the mountain and lowland rain forest types is the scarcity or absence of lianes in the former, the rarity of buttressed trees and the great preponderance of evergreens. The abundance of epiphytes in the mountain forest increases with altitude. The shrub layer throughout is dense but again the herb layer may be sparse if the canopy is continuous.

The distribution of mountain rain forest has also been much altered by cultivation. The soil at the higher altitude appears to tolerate more prolonged cultivation without the drastic loss of fertility found after a few years in the lowlands. The tendency is therefore towards more settled patterns of agriculture, but if a cleared area is abandoned, secondary succession follows a pattern similar to that of lowland rain forest.

Bamboo forest (see Plate 11) may form a distinct vegetation zone between 2,500 and 3,100 metres. Thus, where it occurs, it lies at the upper limit of the mountain rain forest. The character of the bamboo, *Arundinaria alpina*, varies greatly. It may occur, especially at its lower altitudinal limit, in almost monotypic stands. The tall, vertical stems may be eight centimetres in diameter and widely spaced so that progress on foot is easy. Vines are scarce and ground cover is limited to occasional patches of nettles or other herb species. Alternatively, on broken terrain or at its upper altitudinal limits, the bamboo may be less than six metres in height. The stems are closely intertwined and vines are abundant. Typically, these scrubby strands are broken by clearings colonized by shrubs and herbs.

The bamboo is succeeded above its altitudinal limits by the fairytale plant forms for which the mountains of tropical Africa are well known. The *Ericaceae*, or tree heath zone, lies between 2,600 and 3,800 metres (see Plate 13). The lower limits of this zone, particularly in the Virunga volcanoes, the classical home of the mountain gorilla, may be occupied by *Hagenia* or *Hypericum* woodland, between 2,600 and 3,100 metres (see Plate 10). Above about 3,700 metres there is also an alpine zone, where giant lobelias and senecios are found (Plate 12).

Use of varieties of forest by gorillas

As we have already seen, the gorilla has a discontinuous distribution range consisting of many scattered populations. Within the range of each population, however, the gorillas show clear preferences for particular habitats. In Rio Muni, for instance, Jones and Sabater Pi found that 'gorillas rested, slept and fed in regenerating forests, especially in areas such as old fields or along roads'. Indeed, they were able to find gorillas by looking for sleeping and feeding areas in forests bordering the roads. This may seem an unlikely method of locating an animal which usually shuns open spaces and contact with humans. Schaller, however, also records of eastern gorillas:

I found little gorilla sign in old secondary or primary forests which support only a sparse herb stratum. Nearly all the major food plants ... grow most abundantly in cultivated fields and in young secondary forest. Consequently the animals concentrate their activity near roads and villages or, where human activity is absent, in the more open valleys and along river courses.

It appears, therefore, that in lowland areas the gorilla prefers secondary or regenerating forests to primary forest. The chimpanzee and orang-utan, by contrast, are well-adapted to life in primary rain forest, a subject to which I shall return later.

Mountain rain forest is a very favourable habitat for gorillas because it tends to be more open and to contain a denser herbaceous layer than lowland forests. Secondary or regenerating mountain forests are again preferred to areas of primary growth. Bamboo forest is an important vegetation type for gorillas only when the young shoots appear during the wet season. In the Virunga volcanoes, for instance, bamboo shoots form an important item in the gorilla's diet during the rainy period from October to December. In the Mt Tsiaberimu region, where bamboo dominates in those areas not yet deforested by man, Schaller suggests that gorillas probably have to confine their foraging to clearings and small patches of mountain forest.

Tracking gorillas

It is no easy task to find gorillas in the fastnesses they inhabit, for the vegetation is dense, particularly in mountain rain forests, visibility is very poor and often one must use a panga to cut a path. In the Virunga volcanoes, a good view of the terrain can be gained by walking along certain of the mountain ridges, but mist rolls across the slopes almost daily, enveloping the jungle in a damp white shroud. The only sure way to locate gorillas is to search for the trail a group leaves as it wanders through the forest and to track along this in the hope of catching up with the animals.

In areas like the *Hagenia* woodlands of the Virungas, which have a dense herbaceous understorey including many plants eaten by gorillas, the apes often batter a broad path through the vegetation. From the direction in which the flattened plants are pointing, one can tell the route taken by the gorillas; indeed these trails are so distinctive that they cannot be confused with those left by other animals like elephants or buffalo. Along their trails, gorillas often leave food remnants such as piles of wild celery peelings or the split branches of *Vernonia*, as well as segments of dung. If the group crosses soft earth, they will also leave knuckle prints and footmarks. From the freshness of a trail one may ascertain how recently it was made, though appearances can be deceptive, for old trails can look quite fresh with an early dew on them or after one of the frequent rainstorms. Sometimes the gorillas travel without trampling the plants very much and leave no feeding signs or dung. Lone silverback males are particulary difficult to follow, and in areas which have been extensively stamped down by buffalo or elephants it is virtually impossible to track gorillas.

Once a group has been located, extreme caution must be exercised and subtle tactics are necessary, for the animals will always flee when surprised by man. In certain circumstances nothing can induce them to remain; for instance, if they are looked down on from higher ground, they usually move off. Much depends too on their previous experiences when encountering man. Whilst working on Mt Sabinio, Graeme Groom tells me that he tracked gorillas which were terrified of humans, undoubtedly as a result of incursions by cattlemen and poachers into this area of the National Park. Groups usually fled at once and the silverback could be heard screaming for some time as he led the animals away. By contrast, on Mt Mikeno, where the park is much better protected, gorillas were sometimes bold and curious. On a first meeting with one large group, we imitated gorilla feeding and chest-beating behaviour, a method of gaining acceptance which was first developed by Dian Fossey. After a few minutes, the gorillas hesitantly approached, some climbing into the surrounding trees to obtain a better view, others peeping out from the dense vegetation fifteen metres away. These were unhabituated animals which had probably not seen human beings in years. Only the silverback male remained out of sight, occasionally barking, beating his chest, and making sudden rushes through the vegetation. As discussed previously, such males will charge an intruder but this is a bluff manoeuvre; there is little likelihood of attack provided the observer keeps still.

When unafraid, gorillas can be very tolerant of human proximity. For instance, in the Virungas, one gorilla group frequently fed and nested next to a plantation where natives were working. It seems that when unmolested by man, gorillas can learn to ignore him. It is this fact which made it possible for fieldworkers to gather much of the information reviewed in the remaining sections.

The sizes of gorilla groups

It is difficult to assess the size of a gorilla group accurately, because all the members are rarely in view at any one time. Since gorillas build nests at night, however, useful information can be obtained by counting the night nests. Some information on group sizes is shown in Table 7. Western lowland gorillas tend to live in small groups of from two to twelve, with six to eight being the average number. These figures refer to work by Jones and Sabater Pi in Rio Muni and to some observations by Merfield and Miller in Cameroun. The habitat in Rio Muni has been disturbed, by agriculturalists and timber companies, whilst gorillas are frequently hunted in some areas. Population densities are said to vary from 0.58 to 0.86 gorillas per square kilometre; in those areas under greatest ecological pressure population densities and group sizes are lowest. The figures for group sizes in Cameroun may be more typical, since Merfield hunted there in the days when there was much less disturbance of the gorilla's habitat. Merfield and Miller recorded group sizes of from two to ten, but only four groups were counted. Recently Julie Webb has begun a much-needed field study of gorillas in Cameroun and her preliminary report contains information on seven gorilla groups, ranging from two to eight, with an average size of four.

There are many estimates of group sizes for the eastern lowland and mountain gorilla (see Table 7) and it appears that in both these subspecies group sizes may be much larger than for the western variety. Thus groups average between ten and eleven members, and it is not uncommon to find over twenty gorillas in a group. In the Virunga volcanoes, mountain gorilla groups vary a great deal depending on the region considered. In his year at Kabara, which is in the western portion of the National Park (see Fig. 52), Schaller encountered ten groups ranging in size from five to twenty-seven with a mean of 16.9 animals per group. He calculated the population density in the Virungas as a whole to be three animals per square mile. During an expedition to census gorillas in 1973 we counted thirteen groups in approximately the same area. The largest group consisted of twenty-one gorillas and the average group size was 10.3. The position in the eastern portion of the National Park, in the area encompassed by the mountains of Sabinio, Gahinga and Muhavura, is very different, for groups are much smaller there. From reports by Donsithorpe, Bolwig, Kawai and Mizuhara, Schaller concluded that there were five or six groups in the area, with an average group size of seven or eight. However, this calculation included one group of eighteen gorillas, and more recent observations by Groom and

Table 7. Estimates of group size for the three subspecies of *Gorilla gorilla*

Subspecies	Location	Number of groups	Range of group sizes	Average group size	Source
Gorilla g. gorilla	Rio Muni	13	2–12	6.8	Jones and Sabater Pi (1971)
Gorilla g. gorilla	Cameroun	4	2–10	6.2	Merfield and Miller (1956)
Gorilla g. gorilla	Cameroun	7	2–8	4.0	Webb (1974)
Gorilla g. graueri	Kayonza Forest Mt Tsiaberimu and Utu region	20	2–25	10.8	Pitman (1942) Frechkop (1953) Kawai and Mizuhara (1959) Schaller (1963)
Gorilla g. beringei	Virunga volcanoes (Mt Muhavura and Mt Sabinio)	10	2–9	4.1	Groom (1973)
Gorilla g. beringei	Virunga volcanoes (Mt Mikeno)	13	2–21	10.3	Groom, Boesch, Brown, Foulds and Dixson (unpublished data, 1973)

Harcourt in the same area of the Virungas indicate that such a large group size is most unlikely. If this estimate is ignored then the mean of the groups reported by other authors is between five and six. In 1972, Groom encountered ten groups in the eastern part of the Virungas, varying in size from two to nine animals with an average size of 4.1.

Group sizes of the mountain gorilla in the western part of the Virunga volcanoes (the Kabara area) are therefore much larger than on the three eastern volcanoes. Further, we found that the proportion of immature (infant and juvenile) gorillas was higher at Kabara (4.3 per mean group size of 10.3) than in the eastern area (1.2 per mean group size of 4.1). Immature animals constitute only twenty-seven per cent of the population of the eastern volcanoes whereas they make up forty per cent of the Kabara population. In this case it is probably valid to take the proportion of immature animals as a measure of the viability of the population. Hence the population on the three eastern volcanoes is less viable than in the Kabara area.

The differences between the two areas cannot be explained solely on the basis of food availability. At Kabara the dominant vegetation zone is *Hagenia* woodland (see Plate 10), ideal gorilla habitat consisting of old, gnarled *Hagenia abyssinica* trees, an understorey of shrubs and luxuriant herbacious ground cover. Bamboo forest survives in a relatively thin, patchy belt and below this is the zone which Schaller called 'dry colonizing forest'. He thought that gorillas did not range into such forests, but in 1973 the census expedition found gorillas in this zone. The eastern Virungas also provide a favourable habitat for gorillas. Abundant forage is available on the lush herbaceous slopes of the ravines on Mt Sabinio, in the *Hypericum* woodland on the summit of Mt Gahinga and on the western slopes of Mt Muhavura. Mountain rain forest occurs on the lower slopes of all three volcanoes and bamboo forest is abundant, particularly in the saddle areas, so that bamboo shoots are available during the rainy season. However, human interference inside the National Park is much greater in the eastern Virungas than in the Kabara area. Herds of cattle graze inside the park, destroying the gorilla's habitat and frightening the animals. Poaching is rife and frequently takes the form of lines of beaters with dogs scaring game towards the waiting hunters. The gorillas themselves are rarely hunted but they tend, obviously, to move out of areas where poachers operate. Thus on Mt Muhavura there were only thirteen mountain gorillas left in 1972, whereas over a hundred gorillas occurred on Mt Mikeno in the better protected western portion of the park.

I noted earlier that western lowland gorillas live in smaller groups than do the two eastern subspecies. Most of the information on western lowland gorillas comes from areas where their habitat is disturbed, however, and where they are hunted. I hope that field studies will be carried out in one of the few remaining undisturbed areas, such as at Mouloundou in Cameroun, or in the Gabon. A realistic comparison with the work done on the two eastern subspecies should then be possible.

Group composition and the formation of new groups

Gorillas live in fairly stable social groups but the composition of these is not static. Changes take place, not only because of births and deaths within groups but also because certain animals may leave or transfer between groups. Schaller found that in ten mountain gorilla groups there were, on average, 1.7 silverbacked males, 1.5 blackbacked males, 6.2 females, 2.9 juveniles and 4.6 infants. Silverbacked males and the younger blackbacks therefore made up only 18.9% of the group, whereas females made up 36.7%. Most of these females were probably fully adult though it is difficult to be sure about this unless a female is accompanied by an offspring.

Since Carpenter carried out his field studies of howler, spider and rhesus monkeys in the 1930s and 1940s, it has become clear that adult females outnumber adult males in the social groups of many primate species. Hence the gorilla is not exceptional in this respect. Indeed in most gorilla groups there is usually only one silverback present at any one time, though Schaller saw up to four silverbacks in large groups of mountain gorillas. Groom recorded seven groups of from three to nine gorillas on Mt Sabinio and Mt Muhavura. In six of these small groups there was only one silverback and only the largest group contained two. Western lowland gorillas also live in comparatively small groups and the evidence available indicates that only one silverback is present in each group.

Since the sexes are born in almost equal numbers, why do adult females outnumber adult males in groups of gorillas? Part of the explanation may be due to differences in mortality rates between the sexes. Most important, however, is the fact that silverbacked and blackbacked males may leave groups and live, for a time at least, as 'lone males'. Occasionally two males may be seen together and Fossey records one unusual group of mountain gorillas which consisted of five males of various ages. Young male gorillas may leave their natal groups as they attain maturity and set up their own home ranges. Schaller found that lone males may enter and leave established groups and he believed that these males were familiar with particular groups in their area and knew whether or not an approach would be tolerated.

If females were to leave established groups to accompany a lone male then this would provide a mechanism for the formation of new gorilla groups and for the prevention of inbreeding. Recently Harcourt, Stewart and Fossey have documented a number of examples of female 'transfers' among mountain gorillas. Females may transfer from their natal groups, either to join lone males and hence form new groups, or to enter an established group. Blackbacked or silverbacked males may 'emigrate' from groups but they do not transfer during inter-group encounters as females do. Male emigration is a common feature of many primate societies, but female transfer is a rare phenomenon. Marsh, for instance, has recently described its occurrence in an isolated population of red colobus monkeys, living in Kenya in forests bordering the Tana River. Harcourt et al. point out several possible reasons why gorillas behave in this way. There is, for instance, evidence that the presence of a leader silverback in a gorilla group may inhibit other males from mating. If inbreeding is not to occur, therefore, the female offspring must transfer to another group or join a lone male. Apparently females transfer actively soon after reaching sexual maturity; they are not 'kidnapped' by other groups or lone males and nor does the resident male prevent them from leaving. He may, however, behave aggressively towards non-resident males. Harcourt et al. record that ten out of twelve incidences of female transfer 'involved intense displays and sometimes fights between the males'. These are intriguing observations but it should be kept in mind that, like Marsh's observations on red colobus monkeys, they refer to one isolated population. Whether they apply to the species as a whole is not known.

Mortality and morbidity

Many factors may operate to limit the size and composition of gorilla populations. Availability of suitable habitat, seasonal changes in food items and factors controlling reproduction are obviously important but here I shall consider predation, disease and accident as factors which control the numbers of gorillas.

Man is the only major predator of the gorilla. In areas where gorillas have recently become extinct, this has occurred due to hunting and deforestation by man. Silverbacked male gorillas are such formidable animals that one suspects few carnivores would risk attacking a group of gorillas. Indeed this may have been one factor which led to evolution of such massive males although other pressures, such as intermale competition for mates, cannot be discounted. Leopards may occasionally stalk gorillas, but it is probable that they only kill the occasional youngster. Leopards are, of course, much rarer in the Virunga volcanoes now than in times gone by; hunters have seen to that. It is possible that during some earlier phase of its evolution carnivores constituted a greater threat to the gorilla than is the case nowadays.

Disease and accident are probably the major causes of death among gorillas, apart from human interference. The same is probably true of many other primates. Fieldworkers have reported relatively few predators of monkeys and apes. There is no shortage of diseases and parasites which afflict primates, however, and there is ample evidence that serious accidents occur among free-ranging apes. In Chapter 4, the occurrence of helminth worms and malaria parasites in gorillas was discussed, and the fact that many parasitic species found in gorillas are closely related to those which occur in man. This is not surprising in view of the close phylogenetic relationship between man and the apes. Diseases which afflict man can often be transmitted to apes and vice-versa. For example, Van Lawick-Goodall reported a polio outbreak which killed six wild chimpanzees and partially para- lysed nine others. The chimpanzees probably contracted the disease from humans in a nearby village. Among wild gorillas, pneumonia and gastro-enteritis have occasionally proved fatal and conditions resembling yaws, leprosy and 'gundu' (a bone disease) have been reported. Fred Merfield, the gorilla hunter, quite often noticed signs of disease and injury in gorillas which he had shot in Cameroun. The following are extracts from his field notes, which are now kept in the British Museum of Natural History:

October 27th 1932, a female gorilla: 'left eye looks blind and right one going? Right hand deformed and very bad sores between the fingers ... Sexual organs and surrounding parts diseased'

November 11th 1933, a female gorilla: 'hand bent inwards and arm could not be straightened ... sore inside left thigh. Right foot crippled and big sore on outside of ankle'

In wild mountain gorillas, Schaller found eggs of the hookworm (*Necator americanus*) in over fifty per cent of faecal samples. Heavy infestations of hookworm are debilitating to humans and the same may be true of gorillas. It is possible, however, that such parasites do not severely affect gorillas except when injury, old age or stressful conditions weaken the animals. For instance, Richard Fiennes in his *Zoonoses of Primates* records the opinion of Sabater Pi that in young, western lowland gorillas captured in Rio Muni, eighty per cent of deaths were caused by infestations of a nematode called the 'nodular worm'.

Turning to accidental injuries as a cause of mortality it is interesting to consider Professor Schultz's observations on the frequency of healed fractures in ape skeletons. The arboreal Asiatic apes exhibit a remarkably high frequency of healed fractures. Thirty-three per cent of gibbon skeletons and thirty-four per cent of orang-utan skeletons examined by Schultz showed one or more healed fractures. Unlike orang-utans, which are slower and more cautious climbers, gibbons are marvellous acrobats, yet it seems they sometimes fall out of trees. Chimpanzee and gorilla skeletons also showed signs of fractures and healing in eighteen per cent and twenty-one per cent of cases respectively. It requires little imagination to suggest that some fractures may be so severe that they prove fatal. Rijsken has described one wild female orang-utan which survived for two months after being paralysed from the waist downwards, presumably as a result of a fall. A similar observation on a wild gorilla was reported by Malbrant and Maclatchy. Gorillas are mainly terrestrial, but juveniles and young adults not infrequently climb trees, so one might expect them to be more prone to accidents. In mountainous areas such as the Virunga volcanoes the general terrain is hazardous and some of the routes taken by gorillas across the mountain slopes are very hazardous. Schaller saw one female mountain gorilla with a broken jaw and other animals with minor injuries possibly caused by falls.

Infant gorillas are more likely to die from injury or disease than older animals are. In 1973 the census expedition found two dead baby mountain gorillas during two months spent in the Kabara area. One was an aborted male foetus and the other was a male infant of less than one year of age. During his year at Kabara, Schaller noted thirteen gorilla births. One of these infants died, one disappeared and a third was so badly bitten that it probably did not survive. Harcourt, Stewart and Fossey have made the interesting observation that two infants were killed during intergroup encounters among mountain gorillas; an unexpected cause of infant mortality.

Schaller calculated that forty-seven per cent of male gorillas probably die before the age of six. This must be an approximate estimate, but it is not unexpectedly high, for high mortality rates during infancy are typical of many animals. Meadow and Smithnells, for instance, have calculated annual mortality rates among human infants born in England and Wales, basing their calculations upon figures published by the Registrar General. They found that in 1971 death rates per million were 19,861 for males and 15,117 for females during their first year. Between the ages of one and four years, however, death rates for males had dropped to 764 per million and, for females, to

638 per million. It seems that deaths occur most frequently during the first year of life and that male infants of all ages experience a higher mortality rate than females. The major causes of death, incidentally, were accidents, respiratory diseases and congenital anomalies, all of which were more frequent in males than females.

There is not sufficient information to determine whether a sex difference in mortality rate exists among infant gorillas, but this seems a reasonable suggestion. For even if lone males are taken into account, there are still more adult females than adult males in a gorilla population. In some species of monkeys, males have a higher mortality rate than females during adolescence as well. Again it is not known if this is the case in gorillas.

In captivity gorillas may live for over thirty years but aged individuals must face considerable problems in the wild. Arthritis and dental decay are quite common among older gorillas. Schultz found that arthritis of the hip joints, lumber vertebral column and mandibular joints is particularly frequent and that 59.6 per cent of museum specimens of gorillas exhibit abscesses of the teeth. Sinus infections, causing skull deformities, are also 'shockingly frequent among old gorillas'. In captivity elderly gorillas such as 'Guy' at the London Zoo may have diseased teeth removed and may be given a special diet. In the wild, however, old animals must find it difficult to keep up with their groups or to masticate large quantities of vegetation.

Home ranges and core areas

In an earlier section on group sizes, I discussed the differences between mountain gorillas in the two areas of the Virunga volcanoes – the eastern population where small groups are under heavy pressure form human disturbance, and the larger population at Kabara where, in the past, human disturbance has been minimal. Why do the groups in the eastern Virungas stay where they are? If gorillas are totally nomadic then one would expect them to migrate to a more favourable area. This problem leads to a consideration of the factors which determine the ranging behaviour of gorillas.

In 1943, a paper was published by W. H. Burt, in which he distinguished between the concepts of 'home ranges' and of 'territories'. He defined territories as areas which animals defend against other members of their own species, whereas home ranges were not defended. Gorillas do confine their activities to particular, familiar areas of forest but they are not generally thought to defend such areas. Hence it is appropriate to refer to an area used by a gorilla group as its home range.

Estimates of the size of gorilla home ranges vary considerably. In Rio Muni, Jones and Sabater Pi found that for six groups of western lowland gorillas, home ranges varied from two to twelve square kilometres (mean 6.12 square kilometres). They report that these home ranges did not overlap, although since groups sometimes 'intermingled temporarily' with one another or with lone males some degree of overlap between their home ranges would seem likely. The various studies of Schaller, Fossey and Elliott on mountain gorillas of the Virunga volcanoes all indicate that home ranges may overlap extensively. However, Alan Goodall's fieldwork on the mountain gorillas of Kahuzi indicates that there is less overlap between home ranges there than in the Virungas. More fieldwork, in areas other than the Virunga volcanoes, will be required to resolve this problem.

Schaller recorded home ranges of between 6.4 and 13.7 square kilometres for six mountain gorilla groups, and he estimated that, in the Kabara area at least, ranges of from sixteen to twenty-four square kilometres might exist. Casimir and Butenandt found that during fourteen months a large group of mountain gorillas in the Kahuzi-Biega National Park ranged through an area covering thirty-one square kilometres. They add, however, that if this group was studied for several years it would probably be found to use a range of between forty and fifty square kilometres.

Recent research by Dian Fossey and Rick Elliott has shown that some mountain gorilla groups occupy quite small home range areas and that many factors interact to control ranging behaviour. In these studies, great care was taken to record the frequency with which groups visited each part of their ranges. Group movements were plotted on maps, using a system of grid squares (16 squares per square kilometre of terrain). A group studied by Elliott contained from three to six gorillas and ranged throughout an area of seven or eight square kilometres, but most contacts with the group occurred within about fifty per cent of this total range. The home range had changed very little in size and position over a period of six years. Dian Fossey found that a neighbouring group which varied in composition from ten to seventeen individuals ranged through an area covering about eight square kilometres.

Lone male gorillas are not nomadic animals, as was once thought to be the case, for they confine their activities to home range areas just as groups of gorillas do. The home ranges of lone silverbacks overlap those of neighbouring groups, so that social contact is possible. This is important, for as we have seen, lone males must be joined by females from other groups in order to start a new group of gorillas. Recently, Tim Caro of the University of St Andrews, has mapped the home ranges of two lone silverback mountain gorillas. He found that one of these males utilized a home range covering 4.4 square kilometres. Caro was only able to observe this male for ninety-one days, however, so it is possible that the animal's home range embraced a larger area than this. Both the silverbacks in this study concentrated their activities in particular 'core areas' of their ranges. They tended to follow circular travel routes when in the core area and to trample the vegetation a good deal. Caro points out that such behaviour may simply be the result of the silverback having to live in a small area. Dian Fossey, however, has observed such 'over use' of the habitat by three lone silverbacks which had recently left their parent groups. She interpreted this behaviour as a method used by lone males to 'defend' their core areas against incursions by neighbouring groups. These observations are very interesting but they refer to a small number of animals. It may be premature to suggest that these lone silverbacks are showing territorial behaviour.

Like lone males, groups of gorillas spend more time in one or more restricted areas of the range than in other parts. These favoured or 'core areas' do not necessarily lie in the heart of the home range, however, neither do they remain static. There are seasonal and other factors which cause gorillas to shift the location of their core areas.

Factors which may influence ranging behaviour

Gorillas show tremendous variability in the distances they travel each day, as can be seen from Fig. 38, which shows distances between nest sites for a group of twenty mountain gorillas studied by Casimir and Butenandt. This group occupied a home range of thirty-one square kilometres, spanning 6.0 kilometres at its narrowest and 8.4 kilometres at its greatest width. The group travelled through this home range at the rate of 600 to 1,100 metres each day. Two exceptionally long 'marches' of between 2,000 and 3,000 metres also occurred, but these have been omitted from Fig. 38. Similar results have been recorded for other gorilla groups, including smaller groups, and also for lone males. It appears that a group of gorillas could wander throughout its entire home range several times each month. This does not usually happen, however, for various factors lead the animals to concentrate their activities in particular parts of their range and to avoid others.

As described in the previous section, a group studied by Dian Fossey occupied a home range of 8.0 square kilometres. Dr Fossey mapped the range on a grid of 130 squares, each of $\frac{1}{16}$ square kilometres and located a core area to one side of the range where she was able to contact the group much more frequently. She defined

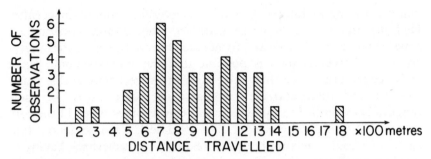

Figure 38. Distances travelled between night nest sites by a group of mountain gorillas. Data are for non-consecutive days. Two exceptionally long daily marches have been omitted. Redrawn from Casimir and Butenandt (1973).

four distinct kinds of travel employed by the gorillas on the basis of how many grid squares of the range they crossed from one day to the next. For a detailed description, the reader will need to consult Miss Fossey's (1974) paper; I shall only summarize her findings here. 'Static' travel involved crossing one grid square, 'regular' travel one to four squares, and 'rapid' travel more than four squares. 'Sudden change' meant that the group abruptly altered direction from a course it had been following for some time. This last type of travel occurred during a period when the group was extending its home range. The animals would move some way into new terrain only to double back and re-enter the familiar range. The type of vegetation in different parts of the range made some difference to the methods of travel employed by the group, but regular travel was the most frequent. Rapid travel was particularly common when the group moved between the slopes of Mt Visoke and the adjacent flat saddle area. These two zones were separated by a transitional area of nettle beds and a broad cattle trail. Gorillas appear to dislike such open areas or parts of the forest where elephant, buffalo or man's cattle have smashed the vegetation. They either avoid these areas or move through them quickly. Fossey discovered also that social factors were important in affecting the group's movements. If it encountered another group near the edge of its range it often responded by showing rapid travel as if to avoid contact. Such a response was less likely, however, if the encounter occurred at the centre of the range or in the core area.

Gorillas spend much of their waking day eating or searching for food. Obviously, attempts have been made to relate ranging behaviour to nutritional factors. This is a complex subject, however, for gorillas feed on many species of plants and to determine the importance and seasonal availability of all of them would be a mammoth task. Schaller recorded a hundred species of plants eaten by the two eastern gorilla subspecies and Casimir found that fifty-six species were consumed in the Kahuzi region. Sabater Pi has listed ninety-two species of plants eaten by western lowland gorillas in Rio Muni. It is very unlikely that these lists are exhaustive. It would be pointless to quote long lists of plant species from the work of these authors, for unless the reader is a botanist, familiar with the areas concerned, then such lists give little impression of the gorilla's diet. However, although gorillas consume large numbers of plant species, only a few of them make up the bulk of the diet in any particular area. Schaller found that seventy-six per cent of mountain gorilla forage consists of vines, herbs and parts of trees. Moreover it is mainly the leaves, bark, stems or roots which are consumed, while the fruits are rarely taken. Schaller considered that, for eastern gorillas as a whole, the major food items are *Aframomum*, *Musa* (banana) and *Urera* (a type of vine), whilst some other plants are important in particular areas. Thus *Mormodica* (a fern) and *Cyathea* (a vine) are often eaten by *G. g. graueri* of the Kayonza forest, whilst *Peucedanum* (wild celery) and *Galium* (a vine) are prominent dietary items for mountain gorillas in the Virungas. A few examples of food items recorded for mountain gorillas are given in Table 8.

There is less information on the western lowland gorilla's diet. Sabater Pi found that leaves, pith and tender shoots constituted fifty-five per cent of the foods eaten by gorillas in Rio Muni. However, he also made the interesting observation that fruits, in the broadest sense, constituted up to forty per cent of the diet. These included not only *Aframomum* fruits but also those of *Gambeya lacourtiana* (the size and colour of an orange), *Grewia coriacea* (cherry-sized reddish fruits) and several others. Differences in the masticatory apparatus between mountain gorillas and western lowland gorillas may, Sabater Pi suggests, be due to the fact that western gorillas have a more frugivorous diet.

Most of the food eaten by gorillas is obtained in areas of secondary forest; seventy-four per cent as opposed to only thirteen and a half per cent in primary forest. In primary forest, the animals forage mainly in groves where herbs, such as *Sarcophrynium*, or vines, such as *Haumannia*, are to be found.

Since gorillas in any particular region concentrate on a small number of staple items for their diet, one would expect the core areas of their ranges to contain abundant supplies of such species and that seasonal changes in abundance might influence ranging behaviour. Two examples will illustrate this fact. Jones and Sabater Pi found that the western lowland gorillas of the Mt Alen area in Rio Muni were most often encountered in forests beside roads or round villages during the rainy seasons. They were able to correlate this with the observation that human activities had created open areas of regenerating forest in these areas and that the gorilla's staple food, *Aframomum*, grows plentifully in such conditions. *Aframomum* produces its fruits during the rainy seasons and these are favoured by gorillas, although they eat the pith of this plant at all times of year. Hence gorillas tend to migrate into the secondary forests around roads and villages in the rainy seasons to feed on *Aframomum* fruit. Jones and Sabater Pi also found that the seeds of *Aframomum* apparently passed unharmed through the gorilla's alimentary tract and germinated in its faeces. Hence the gorilla may act as a dispersal mechanism for the plant.

Turning to eastern gorillas, Michael Casimir and Eckardt Butenandt have produced an elegant study which demonstrates how the seasonal availability of bamboo shoots affects migration and core area shifting in a group of mountain gorillas. Casimir mapped the extent of bamboo forest within the range of a group of twenty mountain gorillas, and showed that bamboo shoots were only available during the rainy season in October and November. When the positions of the gorillas' nest sites were plotted on the map, it was found that during two successive years the group migrated and shifted its core area into bamboo forest during the rainy season (Fig. 39). The gorillas fed intensively on bamboo shoots during October, but when the shoots began to turn woody in November and December, the gorillas migrated back into areas of secondary forest. Since bamboo forest is found in many areas inhabited by gorillas, it seems reasonable to assume that the animals would make annual migrations similar to the one Casimir and Butenandt observed.

Since the home ranges of gorilla groups or of lone males overlap, neighbouring animals may meet from time to time. It is worth considering how such interactions might influence ranging behaviour. The evidence of many thousands of hours of observations on mountain gorillas indicates that interactions between groups are infrequent. Schaller records that, on twelve occasions when groups which he was watching heard other gorillas in the distance, they made no attempt either to approach or avoid them. Fossey, on the other hand, found that the silverbacks in certain gorilla groups sometimes exchanged hooting vocalizations and chest-beats when they were up to one kilometre apart. It is possible, at least in particular cases, that these displays serve to maintain or increase the distance between gorilla groups.

When groups do meet they may apparently ignore each other, or intermingle for varying periods without any signs of aggression. In

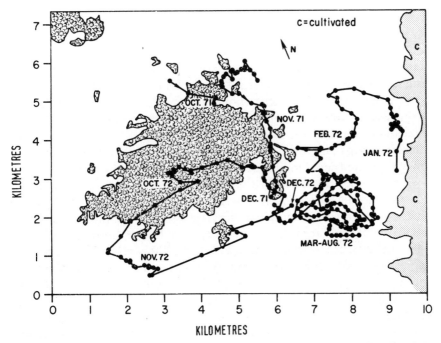

Figure 39. Annual migration and core area shifting by a group of mountain gorillas. Stippled area is bamboo forest; closed circles are positions of the group's night nesting sites. For explanation see text. Redrawn from Casimir and Butenandt (1973).

a few cases, aggressive interactions involving groups or lone males have been observed. As mentioned previously, Harcourt et al. recorded aggressive interactions between males during ten out of a dozen instances of female transfers. However, Schaller recorded an encounter between two groups during which they rested and fed in full view of one another. The silverback in one group bluff-charged at his opposite number and 'the two males then stared at each other, sometimes with brow ridges almost touching'. On three other occasions groups met, intermingled or slept in the same nest site, apparently without any aggressive exchanges.

During his three-month study of lone silverbacked mountain gorillas, Caro observed one male which interacted agonistically with a lone silverback and with three other groups whose ranges overlapped his own. Most interactions occurred in or near the core area of the male's range and involved displays such as hooting and chest-beating rather than overt aggression. Peaceful interactions were also recorded, and on one occasion the silverback fed within fifty metres of his parent group. Elliott watched a group of mountain gorillas for five months and saw only eight encounters with other groups. When one silverback met the group, he fled for one kilometre and 'gave mild alarm vocalizations for over an hour'. The group itself seemed unaffected by this encounter. Alan Goodall observed that a group of mountain gorillas at Kahuzi moved directly from one side of its home range to the other after a contact with a lone silverback.

It would seem then, that mountain gorilla groups rarely behave aggressively towards one another, except perhaps during female transfers. Some of the above examples indicate, however, that particular groups and lone males may have bad social relationships and may regulate their ranging behaviour in consequence of this.

The studies discussed in this section have provided some clues as to why gorillas may journey large distances and occupy extensive home range areas. As noted earlier, it seems that the mountain gorilla tends to live in larger groups and occupy bigger home ranges than does the western lowland form. This is probably related to the greater abundance of forage in mountain forests. Food is usually not in short supply, but seasonal changes in its availability do provide a partial explanation for the gorilla's ranging behaviour. Why do the animals not stay in one locality and exhaust all the forage before moving on? It has been suggested that gorillas cannot obtain all the plants necessary for a balanced diet in any one area, though one feels that the *Hagenia* woodlands of the Virungas probably contain everything that they need. More pertinent is the suggestion that overuse of any one area might cause its destruction. Some years ago domestic cattle almost ruined the area around Kabara in the Virungas, due to overgrazing. A static group of gorillas might also create a great deal of damage. Vesey-Fitzgerald has shown that African buffalo regulate the frequency with which they graze in different areas, allowing sufficient time for plants to regenerate between visits. The same argument might also apply to ranging behaviour in gorillas. It is also possible that long-term changes in climate, resulting in vegetational changes, are best faced by an animal with a large range, which can explore new terrain as the need arises. Climatic changes may have played an important role in determining the present-day distribution of the gorilla, as was discussed in Chapter 2. Finally we have seen that social factors too may play an important role in modifying the ranging behaviour of some gorilla groups and of lone males.

Relationships between gorillas and other mammals

Except for man, the gorilla's relationships with other mammals seem to be mainly placid and non-competitive. Most species are ignored, but gorillas do seem to avoid areas frequented by elephant or buffalo.

This is probably because such large herbivores may damage the vegetation or consume plants the gorillas would eat. Jones and Sabater Pi found that western gorillas did not interact noticeably with other primates such as *Cercopithecus* monkeys or mandrills, which lived in the same area. One crucial question concerns the ecological relationship between the gorilla and its closest relative, the chimpanzee. Whereas gorillas favour secondary forest and feed mostly at, or near, ground level on leaves, pith, shoots, roots, etc., chimpanzees prefer primary forests where they climb a great deal and feed principally upon fruits. Hladik has found that the diet of chimpanzees in Gabon consists of sixty-eight per cent fruits, twenty-eight per cent leaves and four per cent animal food. 'Fruits form up to ninety per cent of the daily intake and never less than forty per cent.' Richard Wrangham's studies at the Gombe Stream Reserve make it clear that feeding is the principal arboreal activity in chimpanzees, whereas resting, grooming and travelling occur mainly on the ground, particularly in the dry seasons. During wet seasons, chimpanzees may spend rather more time resting or grooming in the trees than at other times of year. Chimpanzees and gorillas therefore do not compete to any marked degree for food or space. Hence they seldom interact with each other, even when they inhabit the same general area. In fact, Jones and Sabater Pi saw gorillas meet chimpanzees only once during their field study in Rio Muni. No interactions occurred and both groups moved away in opposite directions.

Daily activities in gorilla groups

Gorillas are diurnal animals and they become active during the first hour after dawn, which breaks at about 6 a.m. in equatorial latitudes. Weather conditions do not greatly influence the time of day at which gorillas rise, although it has been reported that on sunny mornings, or following a night of heavy rain, the animals stay longer at their nest sites. Mountain gorillas and western lowland gorillas have a daily activity rhythm similar to that of many other primates, including gibbons, talapoin monkeys and chimpanzees. As can be seen in Fig. 40, after rising the group feeds, but this activity wanes as the morning progresses. Between 10 a.m. and 2 p.m. most of the group usually rests. During this long mid-day rest period, the group often beats down a large area of vegetation and the animals gather together. Schaller observed that they tended to sit near to the 'dominant' silverbacked male. A few animals may construct day nests, others continue to feed intermittently or sit quietly and sleep. Juveniles and infants play together, whilst some animals groom themselves or each other.

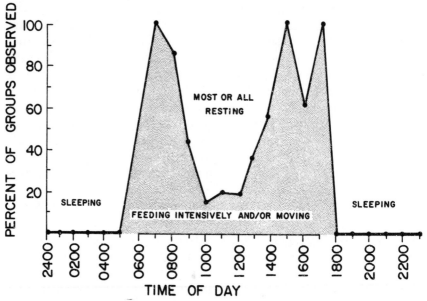

Figure 40. Daily activity rhythm in groups of mountain gorillas. Redrawn from Schaller (1963).

It is possible, however, to watch such a resting group for long periods and yet see little or no social interactions. The afternoon is devoted to more feeding with occasional spells of rest; a gorilla group usually travels further in the afternoon than it does during the morning. Finally, the animals construct their night nests just before darkness falls, which is at about 6 p.m. As one might expect, the cycle of daily activities is influenced by a number of factors, including weather conditions. Gorillas cease feeding during heavy rain and often sit in the open, hunched over with arms folded, enduring a thorough soaking. When there is sunny weather, the gorillas seem to enjoy this. They have been observed sunbathing for long periods until their faces are beaded with sweat.

When a group of gorillas is feeding, the animals spread out and either snatch at plants while still on the move or settle back on their haunches, reaching out in all directions with their enormous arms. During bouts of intensive feeding, a gorilla stuffs food into its mouth with one hand, whilst already reaching for the next fistful. In 1886, Hartmann, citing Koppenfels as his authority, gave the following erroneous but charming description of how gorillas feed:

the head of the family remained at his ease, while his wife and family plucked fruits for him ... and if they were not sufficiently nimble, or if they took too large a share for themselves, the old gorilla growled furiously and inflicted a box on the ear.

In fact each gorilla usually collects and consumes its own forage. Sabater Pi has seen young western lowland gorillas breaking off branches bearing the fruits of *Gambeya*, *Grewia* and *Antrocaryon* and then throwing them down to other group members. These are unusual observations. Others have observed that young gorillas climb and feed in trees more than adults, and that adult females climb more frequently than the enormous silverbacked males (see Plate 14). These age/sex differences in feeding behaviour may be important in reducing competition between group members. Since a group of gorillas usually spreads out as it feeds, and since its members may show individual preferences for particular food plants, competition for available resources is reduced.

Beyond the occasional slug or insect, gorillas do not consume animal matter. They rarely drink water, which may seem surprising until the moisture content of their food is taken into account. Alan Goodall has calculated that a 200-kilogramme silverbacked male gorilla eats approximately thirty kilogrammes of vegetation each day. This contains from twenty-four to twenty-seven litres of water. Allowing ten litres of water lost in the animal's dung and urine or by evaporation, this leaves at least fourteen to seventeen litres of water; more than sufficient for the male's needs.

Gorillas do not simply grab any vegetation; they are highly selective feeders and consume only certain parts of particular species (Table 8). For instance, when mountain gorillas feed on *Rumex ruwenzoriensis*, they uproot the whole plant but eat only the inside of the lower part of the stem and discard the rest. Only the tap roots of *Cynoglossum* are taken: the gorillas either dig them up with their hands or work them loose, shaking them to remove the soil. The pith of wild celery (*Peucedanum linderi*) is preferred and gorillas carefully remove the leaves and outer coating of the stem before eating this. The thistle, *Carduus afromontanus* is, surprisingly, a common food item of the mountain gorilla and is dealt with in an unconcerned fashion. The gorilla runs its clenched fingers up the petiole from base to tip and eats the material stripped off, including the prickles. The petiole is sometimes consumed. Stinging nettles of the most virulent kind (*Laportea alatipes*) are dispatched in an equally uncompromising fashion. Whatever limits the choice of food items, it is clearly not the presence of stings or prickles which can penetrate several layers of clothing. Some plants, however, do present special problems. The vine *Galium* is a favourite of gorillas in the Virungas but its straggling stems and small leaves, armed with hooks, are hard to manage. Adults deftly roll the vine into a ball and then eat it; juveniles may have more difficulty, for it apparently requires practice to perfect these manoeuvres. An example of how deftly gorillas may prepare certain food items is shown in Plate 15 which shows an animal carefully removing and eating the bark of a vine (*Urera hypsolendron*).

It seems that feeding preferences and techniques are learned activities and that young gorillas learn by watching their elders. Gorilla mothers do not encourage their offspring to eat certain plants but they do discourage them from eating species with which they themselves are unfamiliar. The repertoire of gorilla food items differs between groups, for instance between gorillas in the Virungas or at Kahuzi, where 'cultural differences' have been discovered. By this it

Table 8. Examples of plants eaten by mountain gorillas (Kabara area, Virunga volcanoes)

Latin name	Common name	Type of plant	Main parts eaten
Galium simense	Galium	Vine	Entire plant
Vernonia adolfi-frederici	Vernonia	Shrub	Pith and flowers
Carduus afromontanus	Thistle	Herb	Leaf, stem
Laportea altipes	Stinging nettle	Herb	Leaf, stem, bark
Peucedanum linderi	Wild celery	Herb	Stem (bark removed)
Rumex ruwenzoriensis	Rumex	Herb	Stem (lower part), root
Polypodium	Polypodium	Fern	Entire plant
Hypericum lanceolatum	Hypericum	Tree	Wood (rotten), bark
Hagenia abyssinica	Hagenia	Tree	Bark

The first five species listed are major food items; the rest are less important. Data are from the Author's observations. For much more extensive information, see Schaller (1963), Fossey and Harcourt (1977).

is meant that gorillas in one area have not learned to eat a certain plant even though it is available and is utilized by groups in another locality. Thus Casimir spent fifteen months watching a group of mountain gorillas at Kahuzi and never saw them eat *Galium* even though this vine is readily available and is a major dietary item for mountain gorillas of the Virunga volcanoes. Alan Goodall has observed the Kahuzi gorillas eating *Galium*, but only very infrequently.

Schaller identified thirty-eight species of gorilla food plants which occurred both in the eastern Virungas and in the Kayonza forest. Only 14.9 per cent of these species were eaten in both areas although, until recent times, they were connected by forest. Cultural differences in feeding behaviour may, it seems, develop quite rapidly. The reasons for such cultural differences are obscure, but similar findings were made many years ago by Itani for Japanese macaques and have been noted among chimpanzees by van Lawick-Goodall. One can only speculate that gorilla groups acquire new feeding habits as a result of infants or juveniles testing unfamiliar plants. Little gorillas occasionally try unpalatable lichens and other types never touched by adults. Hence, over the course of time, differences might develop in food choice between groups and between populations of gorillas. Adult gorillas are more inflexible in their feeding preferences. Michael Casimir offered bamboo shoots and two species of vines eaten by mountain gorillas to a captive adult male *G. g. graueri*. The animal sniffed the food items, but repeatedly refused to eat them, for he had been kept in captivity since infancy and was unfamiliar with the plants.

In an attempt to find out why gorillas choose particular parts of plants and discard the others, Casimir has subjected various parts of nine plants eaten by mountain gorillas to chemical analysis. The analyses were for protein content, concentrations of individual amino acids and water content. Information on mineral content was also collected for eight species. Unfortunately, careful though these studies were, no overall correlation between chemical content and food selectivity could be found. Gorillas do not, it appears, select parts of plants which contain the highest concentration of a particular nutrient or combination of nutrients. Casimir points out, however, that concentrations of vitamins and fats were not analysed in these plants. Also parts of certain plants contain poisonous substances and are probably avoided for this reason. Thus mountain gorillas in the Kahuzi region eat only the leaves of *Basella alba* and discard the fruit. The fruits, but not the leaves, of the plant contain a haemolytic poison.

The gorilla's nests

Like all the great apes, the gorilla builds nests and a great deal of information can be obtained about a gorilla group by studying its nesting sites. Nests often remain identifiable for many months and so they are frequently the only sign of the apes which a casual visitor encounters. Gorillas build nests for sleeping in at night and, sometimes, for use during the mid-day rest period. Day nests are recognizable because there is usually no dung in or beside them and because they are not as flattened as night nests, having been occupied for only a brief spell.

Plotting the positions of a group's nest sites on a map can give invaluable information about ranging behaviour (see Fig. 39).

Further, since all the animals in a group over about two and a half years of age build new nests each evening, counting the nests gives a strong indication of group size. It is much easier to count nests than the animals themselves, for only a portion of a group is ever in view at one time. When examining a collection of night nests it is advisable to mark each one with a slip of paper in order not to count it twice, for in dense undergrowth it is easy to make this mistake. It can take more than an hour before one is sure that all the nests have been located, even if they are concentrated in less than six hundred square

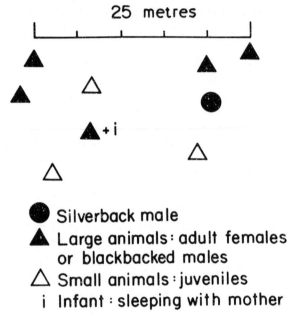

Figure 41. Diagram of a night-nest site used by ten mountain gorillas.

metres of terrain (Fig. 41). The presence of infants or juveniles is often indicated by small segments of dung on the rim of the mother's nest. Nest counts have misled some workers, such as Bingham, who recorded clusters of over thirty nests for the mountain gorilla. What probably occurred was that two groups had slept close to each other, or nests had been built in the same place on two different evenings.

Some clues about the age/sex composition of a group can also be obtained from nest counts. Silverbacks build big nests, five feet or more in diameter, and these are usually distinguishable from those of blackbacked males or females. A still more valuable clue is the size of the dung which gorillas in many areas deposit in their nests during the night. Gorilla dung consists of a chain of segments which vary in size depending upon the animal's age. Silverbacks produce segments seven centimetres or more across whilst those of females or blackbacks are somewhat smaller. Juvenile dung is from two and a half to five centimetres across; that of infants is less than two and a half centimetres. At Kabara, almost every night nest of the mountain gorilla one examines contains dung, but this is not true for all gorilla populations. Casimir has found that adult mountain gorillas at Kahuzi only defaecate in fifteen per cent of their night nests whereas youngsters defaecate in seventy-four per cent of nests. In Rio Muni, Jones and Sabater Pi found that western lowland gorillas leave dung in forty-three per cent of nests whilst dung is left outside a further twenty-six per cent and the remainder contain none. Whether such differences are 'cultural' or merely relate to variations in the bulk and types of food eaten is not known.

The fact that gorillas may defaecate in their nests has led to rather anthropomorphic comments about their 'most unsanitary behaviour', but in fact the droppings are firm, fibrous, and do not foul their pelage noticeably. Vernon Reynolds has made the interesting observation that chimpanzees, which produce soft faeces as a result of their frugivourous diet, are careful to void them outside the nest.

Nesting begins at the end of the afternoon feeding period, up to an hour before dusk. The leader silverbacked male starts his nest, usually near the spot where he has been feeding, and the group members soon follow his lead. The technique used varies with the situation, but there are some constant features. As long ago as 1861 Paul Du Chaillu accurately described nest-building behaviour. The simplest case is the nest built on level ground. The animal bends or breaks the surrounding vegetation, folding it around and under its body to produce a full or half circle of herbage (see Plate 16). The material is usually twisted as it is bent over and this helps to keep it in place. Often the only quality of vegetation which the gorillas require is that it should not snap off completely when bent into the nest. Sometimes a gorilla will reach out and grab vegetation just a few times and complete its nest in under a minute, though Schaller saw one animal spend five minutes on the operation. Most attention is directed at the construction of the nest rim and often the centre of a ground nest is left unattended to, so that the animal rests on the bare ground. When the gorilla sleeps, it lies on its side with legs and arms drawn up, or on its belly, with the limbs tucked underneath it. There is considerable variation as regards the care a group takes in building its nests on different evenings. Occasionally mountain gorillas make no nests, perhaps because the animals have continued feeding until nightfall. When a gorilla sleeps on mountain slopes it may need to construct a platform before making its nest. Bolwig describes gorillas on steep slopes pulling branches down from above them and anchoring the tops in the vegetation below. The animals then climb up onto the platforms they have made. Gorillas sometimes nest in the trees and, as one might expect, they take longer over this, for such nests must be robust enough to support the animal's weight while it sleeps. They usually build where branches fork, but make no attempt to intertwine the foliage and produce a stronger framework.

Chimpanzees and orang-utans also build nests at night. Robert Yerkes considered nest-building to be a genetically predetermined behaviour in chimpanzees but that animals improved their techniques by practice as they grow up. As we saw in the last chapter, captive-born and isolation-reared chimpanzees are inferior in their tool-using abilities when compared with wild-born animals. The same is true as regards nest-building. Irwin Bernstein studied the responses to nesting materials of seven adult, wild-born chimpanzees and eighteen captive-born adults. All the wild-born chimpanzees built nests, but only eight of the captive-born individuals did so and in five cases their attempts were very poor. Bernstein concluded that learning played an important part in the development of nesting patterns. He also found that the chimpanzees used the most suitable materials provided to build their nests, whether these were plants or artificial substances. In the wild, gorillas also exhibit great flexibility when choosing nesting materials; they simply use those plants which are available in a particular locality. At Kabara, for instance, I saw mountain gorilla nests containing senecios, wild celery, vernonias and lobelias, all of which are plentiful in that area. High up on the mountain slopes, however, different plant species occur and gorillas often used giant senecios to build their nests. In Rio Muni, Jones and Sabater Pi found that western lowland gorillas often make their nests in dense stands of their favourite food plant, *Aframomum*.

In the wild, Schaller found that young mountain gorillas sleep with their mothers for the first two and a half to three years, but they start to make 'practice nests' during this time. The earliest instance he observed was an attempt made by an eight-month-old and the first full ground nest was built by an infant of eighteen months. Kawai and Mizuhara compared the young gorilla leaving the maternal nest to the binary fission of an Amoeba:

First the mother's nest swells into a gourd-like shape. In the larger part sleeps the mother and the smaller part is for the baby. Soon the nest separates into two parts, but the baby's nest remains close to the mother's. Then at last, separation is completed and the baby's nest becomes independent.

This delightful idea has not, however, been confirmed. Juveniles do nest close to their mothers when they first sleep on their own, but reports of gourd-shaped nests are very rare.

A number of attempts have been made to relate the position of the night nests built by a gorilla group to certain functional considerations, such as defense against predators. Old reports conjure up quaint visions of the silverback watching over his family at night. Koppenfels asserted that the females and young slept in a tree whilst the big male sat on guard below with his back against the trunk. More recent accounts by Blower, Lequime, and others claim that nests are constructed in a certain pattern so that the gorillas can see each other and the surrounding terrain. There is little evidence for any of these conclusions, for nest sites of particular groups have been mapped over long periods and show no consistent patterns. It is possible that if an observer knew all the animals well enough, and the social relationships between them some order might emerge concerning which apes nest closest together. Age and sex differences might be important in this context. Schaller found that, on an average, 10.4 metres separated nests built by adult males whereas juveniles nested only 3.6 metres apart. Medium sized animals (either blackbacks or females) slept 1.6 metres from each other and 0.9 metres away from juveniles. Unfortunately, more recent research, by Michael Casimir, on a group of mountain gorillas of Kahuzi has failed to confirm Schaller's findings.

Jones and Sabater Pi have made a study of ecological factors which may affect the nesting behaviour of western lowland gorillas. *Aframomum* was included in the construction of ninety-five per cent of nests and *Sarcophrynium* in fifty-four per cent. This may not reflect a preference for these plants as nesting materials, however, but may simply underscore the fact that gorillas feed on them a great deal and build their nests wherever they happen to be foraging at the end of each day. The same study makes further correlations between location of nests and slope exposure or overhead cover. Thus forty-six per cent of nests were on south and east facing slopes. Ninety per cent of all nests lacked any overhead cover. It seems very unlikely, however, that gorillas calculate their day's travel so as to end up on south and east facing slopes. More probably these slopes tend to have a greater abundance of forage for the animals. Since most nests are constructed of *Aframomum* they naturally lack overhead cover, for this low plant requires open situations in which to grow and, once pulled down for building, is bound to leave the nest exposed. Therefore, despite much painstaking observation by more recent workers, Harold Bingham's original conclusion made in 1932, that 'the location of nest sites each night appeared to be a matter of expediency rather than of selective foresight' still stands.

Some confusion has existed in the past as to whether gorillas build their nests on the ground or in the trees. It would be most accurate to say that they do both, but that ground nests are the norm in many areas of the gorilla's distribution range. Mountain gorillas of the Virunga volcanoes construct nearly all their nests on or within a few

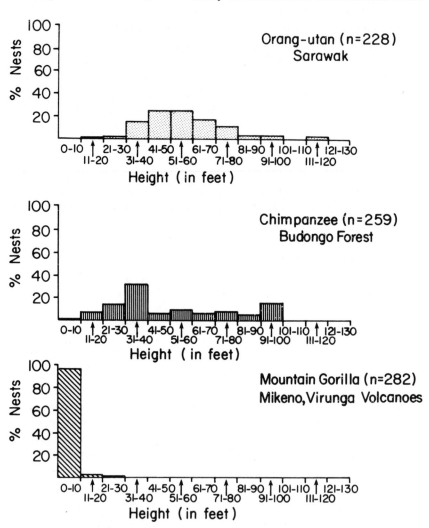

Figure 42. Heights at which the three great apes build their night nests.
Data on orang-utans from Schaller (1961) and on chimpanzees from Reynolds and Reynolds (1965). Data on mountain gorillas of the Virungas are from the author's field notes. Gorillas in some other areas may build a higher proportion of tree nests – see text.

metres of the ground (Fig. 42) and when one climbs up to the tree nests it is usually clear from their size, and the dung inside, that they are the work of smaller animals. A report of two populations of western lowland gorillas notes that eighty-four per cent of nests were at ground level. Eleven per cent were less than two metres above ground, and nearly all the rest were between two and eight metres. In the Utu region in Zaire, however, Schaller found that *G. g. graueri* regularly nested in high trees. They did so with a frequency that was nearly twice that of eastern lowland gorillas in the Kayonza forest, where the vegetation was no less suitable for arboreal nesting than in Utu. He thus tentatively suggested a regional difference in habit. However, since the Utu gorillas are hunted for meat and the Kayonza population is not, harassment might be the ecological determinant encouraging nesting off the ground. Moreover, some regional variations in nesting behaviour are perhaps attributable to differences in the types of vegetation available. Nests built in thickets of bamboo, for instance, are nearly always raised above ground because of the height and density of the vegetation. However, it is possible that in its nesting and feeding behaviour, *G. g. graueri* is more arboreal than *G. g. beringei*. One bears in mind that the latter shows adaptations of its arms and feet which would indicate that it is the most terrestrial of the three gorilla subspecies.

In contrast to the gorilla, the other great apes are primarily tree nesters (see Fig. 42). Indeed, orang-utans rarely come to the ground. Unlike gorillas, orang-utans sometimes use old nests for sleeping in. These apes are heavy animals and they build strong enough nests to provide comfortable platforms in the trees where they may sleep safe from predators. The gibbon and siamang are much smaller apes and do not build nests; instead, they sleep sitting on two hard pads on their rumps: the ischial callosities. As discussed in Chapter 3, ischial callosities are lacking in the great apes, but occur in Old World monkeys. The forerunners of the gorilla were undoubtedly quite large arboreal creatures which built tree nests just as chimpanzees and orang-utans do today. As gorillas took to a terrestrial life and increased in size, they retained the habit of building nests, but on the ground where they are no longer of much value to the animal.

Social structure in the gorilla and other apes

We have seen that a group of gorillas usually contains one silver-backed male, although as many as four have been observed in large bands of mountain gorillas. Some males leave groups and live alone for varying periods, whereas mature females may transfer between groups but never lead a solitary life. This type of social system has many advantages for a creature which reproduces slowly and whose offspring are dependent upon the mother for a long period. A pregnant female, or one with an infant, would be highly vulnerable to predation outside the protection of her group. The group environment is also advantageous because it provides opportunities for young animals to learn and perfect the behavioural skills necessary for survival.

Most anthropoid primates live in social groups, but the structure of these varies tremendously. Nowhere is this variability more apparent than among the apes. It seems that ecological pressures, rather than phylogenetic ones, are the crucial determinants of primate social structure. Gibbons and siamangs, for instance, live in family groups, consisting of an adult pair plus their offspring. Western man might argue anthropocentrically, that this is a 'highly evolved' social system and the more so because it occurs in a closely related primate. However, the *Callicebus* and *Aotus* monkeys of South America also live in family groups, whereas man's closest relatives, the great apes, do not. Gibbons are small, territorial, arboreal apes and, in John Ellefson's view, are adapted for feeding at the ends of branches where bulkier forms could not reach. The orang-utan by contrast is more frugivorous and is so large that it must forage slowly and can only live in a particular type of forest. Adult males are usually seen alone and adult females may be accompanied by an infant or juvenile. This solitary existence, unique among the apes, is a reflection of the animal's great size, arboreal habits and frugivorous diet. Large apes, which must move about the forest in search of fruiting trees without descending to the ground, could scarcely exist in big social groups. The chimpanzee is also a fruit-eating ape but it has a social system and method of foraging quite different from that of the orang-utan. Chimpanzee groups are very large but they split into many small sub-groups whose size and composition frequently change. These sub-groups move around a great deal on the ground but they are able to locate fruiting trees more rapidly than the arboreal orang-utan. When food has been located chimpanzees may pant-hoot and males also drum on the buttress roots of trees. This 'booming clamour', as Sugiyama has called it, tends to attract other sub-groups to the food source. Gorillas, unlike the orang-utan or chimpanzee, feed mostly at ground level on large quantities of fairly un-nutritious herbage. Their forage is more abundant, less localized in its distribution and less subject to seasonal variations than the fruits which chimpanzees feed upon. The relatively stable social groups in which gorillas live are successfully adapted to this particular way of life.

When Schaller observed wild gorillas he attempted to define a 'dominance hierachy' between group members, similar to those recorded by fieldworkers who had studied other primate species. Since aggressive interactions were very infrequent he assigned ranks to animals on the basis of more subtle indicators, such as one gorilla avoiding another or giving way on a narrow trail. Schaller found that when more than one silverback is present in a group, then a linear rank order is apparent amongst them. Silverbacks always rank above adult females and blackbacked males. At the bottom of the hierarchy come first juveniles and then infants. A female with an infant outranks a childless female and blackbacks rank above females in some cases. In captive gorilla groups, such as the one studied by Jorg Hess in Basle Zoo, agressive interactions seem to be more frequent than in the wild. This increased aggressiveness of captive groups compared to their wild counterparts has been found for a number of primates. For instance, Thelma Rowell studied baboons in a Ugandan forest and also in a captive group. She found that the caged animals showed a much higher frequency of aggression and a more rigid hierarchy than the free-ranging baboons did. The abnormal physical environment of a cage, coupled with the fact that captive groups often consist of strangers taken from different groups in the wild, may result in heightened levels of aggressive, sexual and other behaviour patterns.

A few words must also be said here about the 'dominance hierarchy' concept as it has been applied to primate societies. The concept was first developed by Schjelderup-Ebbe who studied 'pecking orders' in groups of domestic fowl. Unfortunately, in the hands of some primatologists, 'dominance' has become a rather nebulous term which has been assessed by a range of criteria including aggressive interactions and priority of access to food, water, grooming and sexual partners. However, as Bernstein and others have shown, for a range of monkey species, if the animals in a group are ranked by several of the above methods, then the resulting hierarchies rarely correlate. A monkey which mates most frequently is not necessarily the one which receives most grooming or ranks highest in aggressive encounters with other group members. When considering Schaller's scheme of hierarchical relationships among wild gorillas the reader should not imagine that it represents a rigid framework in which each gorilla 'dominates' the lower ranking members in a variety of ways. Primate social structures are much more subtle than this. Interaction between group members may be influenced by kinship ties and other factors.

In the past, male gorillas have often been described as very aggressive animals. For instance, in 1928, Chorley recorded of mountain gorillas:

The old grey-backed male seemed to have a surly temper, for sometimes he would grab one of his wives by the head and succeed in throwing her ten yards away. Nevertheless his wives seemed to regard him with real affection...

There is also a certain mythology to suggest that when he grows old, the dominant silverback is driven out by a younger male. The following description, concerning a group of western lowland gorillas, was made by Merfield:

The largest of the two half-grown males had begun to cheek his father. I was sorry, for the old man was an excellent parent and he did not deserve his inevitable fate of being challenged and finally driven out to a lonely old age.

In reality, gorillas lead a placid and unaggressive existence. A few genuine fights between silverbacks have been recorded but these are exceptional and serve to prove the rule of peaceful co-existence between males. Only one silverback acts as leader but his relationship with other males in his group are not antagonistic. Elliott found that two silverbacks in a group of mountain gorillas maintained a greater inter-individual distance than any of the other group members. He points out, however, that one of these silverbacks perhaps preferred to remain in a peripheral position, and did not necessarily do so to avoid aggressive interactions with the leader male.

Without doubt the leader silverback fulfils a most important role in a group of gorillas. He decides which routes are to be travelled, when the group will rest and where it will sleep each evening. During a rest period, Schaller found that if the leader stood stiff-legged and walked off quickly, then the other group members followed him. When a group is moving rapidly, the silverback often walks at its head; the sight of gorillas 'steamrolling' through the undergrowth in this fashion is an unforgettable one (see Plate 17).

Communication by postures and facial expressions

No comprehensive account of how gorillas communicate by means of postures, gestures and facial expressions has yet been published, though there is an extensive literature available on the chimpanzee and some useful information on the orang-utan. The following account is based on the researches of Fossey, Hess, Nadler, Schaller and Yerkes, as well as my own limited observations of wild gorillas or animals kept in zoos. Communication in sexual contexts will be dealt with in the next chapter.

Gorillas are renowned for their extraordinary chest-beating behaviour (see Fig. 14,), but until Emlen and Schaller studied mountain gorillas, it was not realized that chest-beating is only one of a sequence of nine actions which make up the total display. The sequence, in its complete form, is as follows: hooting, symbolic feeding, rising bipedally, throwing vegetation, chest beating, leg-kicking, running, slapping or tearing vegetation and thumping the ground. Only silverbacked males perform the entire sequence, which lasts for about half a minute, but all age and sex classes perform at least some of the above actions. Emlen has pointed out that the display elements are not as rigidly ordered as those which occur in many fish or bird displays. Hence, whilst watching the observer, a gorilla may shake or pound on a branch, or beat its chest, but omit other portions of the display. Only silverbacks hoot at the start of their displays, and sometimes they hoot but then fail to carry on and complete the display sequence. Alternatively, they may proceed to place some vegetation in the mouth and then spit it out. This 'symbolic feeding' may, in ethological parlance, be described as a 'displacement activity'. Thus an apparently irrelevant activity, in this case feeding, has become ritualized and incorporated into the display during the course of its evolution.

Before it beats its chest, a gorilla rises and often pulls up vegetation, throwing it into the air. Then it performs a volley of chest beats using alternate, slightly upward movements of its forearms. Sometimes the animal beats on branches, and captive gorillas will pound the walls of their cages. The usual number of chest beats is about ten, with as many as twenty being given sometimes and towards the end of the sequence a gorilla may slap its belly rather than its chest. The sound produced is difficult to describe: in adult males it resembles a loud 'pock-pocking' which can be heard over long distances, whereas females and blackbacks produce a much duller note. The extensive system of laryngeal diverticula which are best developed in adult males were described in Chapter 3. When dilated, these sacs probably function as resonators, while the hands are cupped to form air traps as the gorilla slaps the naked skin of its chest.

Whilst chest-beating, a male may kick one leg in the air and then run sideways for a short distance, swiping at the vegetation as he does so. Finally he swings one arm and brings the flat of his hand down on the ground with a loud thump. Other gorillas tend to move away from a silverback during his display, for he may blunder into them or strike them during the running and slapping stages. Indeed, Emlen has suggested that the hooting which precedes the male's display may alert other group members and warn them to move out of his way.

Why do gorillas beat their chests? There seem to be many answers to this question. All the members of a group will chest-beat on some occasions and the behaviour seems to be contagious. One animal will beat its chest only to be answered by another some yards away and out of view. I have tried chest-beating at gorillas and have recieved a similar response from them. There was no intimidation in these situations; it is more likely that the gorillas communicate their position to one another by means of their chest-beating. Schaller saw some gorillas chest-beat during play and in situations of sexual excitement. Yerkes observed chest-beating by a captive female mountain gorilla when she was frustrated in her attempts to obtain food rewards during intelligence tests.

Gorillas may chest-beat in response to another group, or to intimidate an intruder. Certainly a silverback's display in response to man serves to advertise his size and strength whilst avoiding an actual attack. Such ritualization of aggression into non-contact displays is a widespread phenomenon among vertebrates. While gorillas are not territorial, Emlen, Schaller and Fossey have all reported that silverbacks in different groups may exchange hoots and chest-beats while some distance apart: whether such displays may function as group spacing mechanisms is not known.

Chest-beating, like nest-building behaviour, is genetically predetermined to the extent that it can occur in individuals who have had no opportunity to learn from other gorillas. This is indicated by observations on infants who were born in captivity, removed from their mothers and hand-raised. For instance the female 'Goma', at Basle Zoo, beat her chest before she was one year old. In wild gorillas the display appears early in ontogeny; Schaller saw four- and five-month-old mountain gorillas beating their chests, and sometimes youngsters did so while playing together. Marcia Stefanick, who studied a group of six juvenile lowland gorillas in Philadelphia Zoo, has kindly told me about her observations. Chest-beating was often directed at another animal and usually some form of interaction would follow between the pair. Often the interaction took the form of play, involving chasing and wrestling.

Huber has shown that the facial musculature and its innervation shows its most complete development in the great apes and is surpassed in this respect only in man. Therefore, one might expect gorillas to make extensive use of facial displays in communication. The brooding and introverted nature of the gorilla's visage has often been remarked upon and its forbidding appearance has given rise to a number of colourful descriptions including the following by Du Chaillu, in 1869, about a male western gorilla:

I could see plainly his ferocious face. It was distorted with rage; his huge teeth were ground against each other, so that we could hear the sounds; the skin of the forehead was drawn forward and back rapidly, which made his hair move up and down and gave a truly devilish expression to the hideous face.

In fact, a gorilla's scalp is hidden beneath dense hair and the ears are small, relatively immobile, and set close to the sides of its head. In many Old World monkeys such as macaques, mangabeys and baboons, scalp and ear movements make a major contribution to facial displays; but they are much less important in the apes. Man still retains some ability to move the scalp and ears, as one can easily demonstrate for oneself, but this is trifling compared to what a macaque or baboon is capable of. Man's brows and forehead are particularly important in facial expression, however, and the African apes also use brow movements in a number of contexts. Van Hooff observed that, like man, but unlike Old World monkeys, gorillas and chimpanzees lower the brows and draw them together in a frown, as an expression of threat. A gorillas's eyes are most expressive, and during threat they change from their normal soft hue and grow hard, staring fixedly at the other animal whilst the head is tilted downward and the lips pursed or parted slightly. During intense aggression, the lips are raised and drawn back to expose the teeth. The gorilla opens its mouth to varying degrees and roars loudly. A frightened gorilla, such as one suddenly encountering man in the wild, opens its mouth very widely and tilts its head back to emit loud screams.

Chimpanzees and gorillas are capable of smiling in the same manner as human beings and in contexts which we might interpret as indicating pleasure (Fig. 43). When young gorillas play and grapple with each other they may open their mouths slightly without exposing their teeth and with the mouth corners drawn back in a smile. Animals may also 'play bite' one another or open the mouth widely in a 'play face'. Such behaviour occurs in many monkey species as well as in the apes.

Marcia Stefanick described a facial expression in captive, western lowland gorillas which she calls 'liptucking'. This occurred in situations where one gorilla was possibly going to interact with a cagemate but seemed hesitant to do so. As can be seen in Fig. 43, one or both lips are folded in against the teeth. Both in its appearance and context this display is very similar to nervous lip-biting in human beings. I have seen this expression in wild gorillas as a response to the presence

Figure 43. Examples of the gorilla's facial expressions. (*Upper left*) a baby gorilla smiling in response to being tickled; (*upper right*) an adult male showing the 'lip-tucking' or 'tight-lipped' face (from a photograph by R.D. Nadler); (*lower left*) a young gorilla shows its tongue (from a photograph by P. Coffey); this expression occurs when gorillas concentrate upon an object or in situations of uncertainty; (*lower right*) an adult male yawning. Author's drawings of captive western lowland gorillas.

of man and Ronald Nadler has also recorded it in captive western gorillas, calling it the 'tight-lipped' face. He found that it nearly always occurred in association with a marked shifting of the animal's gaze away from the receiver of the display. When surprised and uncertain, or when concentrating on some object, gorillas sometimes part their lips and expose the tongue to view. This behaviour was mentioned in Chapter 1 when describing Rupert Garner's attempts to study gorillas in the wild. Tongue-showing also occurs in similar contexts in humans. In situations of 'stress' or 'uncertainty' gorillas sometimes yawn widely. This expression occurs in many monkeys and has been referred to as 'tension yawning'. Tinbergen has noted that humans also yawn more frequently when they are nervous. Some years ago, Devore suggested that yawning could function as a threat in baboons, by displaying the massive canine teeth. As far as I know, however, gorillas do not employ yawning as an aggressive display. In certain species at least, yawning is more frequent in males than in females; indeed Phoenix has shown that in rhesus monkeys, yawning decreases in frequency when adult males are castrated and may be restored by injections of testosterone propionate.

Chimpanzees use some of the same postures, gestures and facial expressions as gorillas. Chimpanzees, however, have a larger and more versatile display repertoire. This is perhaps related to the fact that chimpanzees live in constantly changing sub-groups; what Hans Kummer has called 'the perfect fusion-fission society'. Such a way of life necessitates a good deal of communication between group members, for instance to reduce the likelihood of aggression occurring when sub-groups meet and mingle. Gorillas, with their more compact groups and passive, herbivorous existence, make far less use of visual displays. In fact, I have already mentioned most of the postures which gorillas employ for visual communication. A few examples remain, however, including the 'strutting walk' which was first observed by Schaller. This is an exaggerated type of walking in which the limbs are held stiffly (see Plate 18). Dian Fossey has seen silver-backed mountain gorillas display in this manner by walking parallel to one another. At Yerkes Primate Center, at Twycross and at London Zoo I have observed adult western gorillas of both sexes using strutting displays in a variety of contexts. Sometimes the display was directed at humans and seemed to have an aggressive motivation, but at other times it was difficult to understand why the gorillas behaved in this fashion. Jorg Hess, who observed western gorillas at Basle Zoo, noted that the silverback 'Stefi' sometimes made a strutting run when a female tried to avoid his sexual advances. The male would run stiff-legged past the female and strike out at her with an arm or a leg as he did so. At London Zoo, the silverback 'Guy' sometimes displayed towards people in a similar fashion.

It is necessary to consider tactile communication as well, and in particular 'grooming'. This behaviour is not merely a cleaning activity for, in many species of primates, the amount which any two individuals groom one another is influenced by their kinship ties, social rank, age and sex. Gorillas groom one another less frequently than chimpanzees and many monkey species do. Perhaps this is part of an overall lower frequency of communication in gorilla groups, in which animals spend most of the time feeding, moving or resting and digesting the large quantities of herbage consumed. Gorillas may groom for cleansing purposes and one animal may invite grooming from another by presenting its rump, perhaps because this is an area which a gorilla cannot reach easily during self-grooming.

Besides the communicatory displays which I have described there are some others mentioned by various workers. These include crouching and hunched sitting with the arms folded or held over the head:

this seems to be a submissive posture. Some other behaviours have been described, but it is not clear if these apply to the species as a whole or whether they are the idiosyncracies of a few individuals. Gorillas probably have a larger and more subtle repertoire of visual displays than is realized and much might be learned by studying groups kept in captivity.

Vocal communication

Dian Fossey has identified sixteen vocalizations used by wild mountain gorillas. One sound was made only by a single sick animal, so the repertoire probably consists of fifteen basic sounds. Marler has pointed out that many mammals and birds have a vocal repertoire of between ten and fifteen elements (for example, prairie dog, ten vocalizations; chaffinch, twelve vocalizations). However, in many higher primates, vocalizations are not discrete, but form an intergraded series. In the rhesus monkey, for instance, Thelma Rowell found that nine vocalizations used in aggressive or submissive contexts intergrade with one another along a single continuum. Intergradation of vocalizations also occurs in the chimpanzee and gorilla.

Gorillas and chimpanzees have a broadly similar vocal repertoire (Table 9) but Marler stresses various important differences between the two species. Two aggressive vocalizations used by the gorilla, the roar and the growl, do not have an equivalent in the chimpanzee's repertoire. Further, in gorillas, vocal communication is dominated by the silverbacked males. Silverbacks vocalize more frequently and use a greater range of sounds than other age/sex classes. The only exception to this rule concerns certain sounds used most frequently by infants; whining, crying or chuckling. Some sounds, such as hooting before chest-beating or the aggressive roar, are used almost exclusively by silverbacked males. Some sounds are principally for communication within the group. When gorillas spread out to feed, group members often lose visual contact with each other. They often give deep 'belching' vocalizations, which probably enable the animals to monitor the position of other group members. Chest-beating may serve a similar function. When a group is on the move, gorillas occasionally emit a series of short 'pig-grunts'. Fossey records that pig-grunts are often exchanged when a 'right of way' dispute occurs on a feeding trail. Silverbacks also employ this vocalization to co-ordinate group movements.

Silverbacks use several vocalizations in response to factors outside the group, such as human proximity, or the proximity of other groups or lone males. 'Questioning' or 'hiccup' barking denotes mild alarm

Table 9. Comparison of chimpanzee and gorilla vocalizations

		Chimpanzee	Gorilla	
	Vocalization	Context	Context	Equivalent vocalization
1	Pant-hoot	hearing or rejoining group	inter group communication	hoot series
2	Pant-grunt	subordinate meeting a dominant	mild threat within group	pant series
3	Laughter	play	play	chuckles
4	Squeak	being threatened; submission	infant separated	cries
5	Scream	fleeing; lost; copulating female	aggression within group; copulating female	scream
6	Whimper	begging; parent-infant separation	danger of injury or abandonment	whine
7	Bark	vigorous threat	alerting to mild alarm	hoot bark
			initiating group movement	
			very mild alarm; curiosity	question bark
			very mild alarm; curiosity	hiccup bark
8	Waa bark	threat to other; at distance	see wraagh	
9	Rough grunt	approaching or eating preferred food	feeding; group contentment	belch
10	Pant	copulating male; grooming; meeting another	copulating male	pant
11	Soft grunt	feeding; social excitement	mild aggression in moving group	pig grunt
12	Cough	mild confident threat to subordinate	see pig grunt	
13	Wraa	detection of predator	sudden alarming situation	wraagh
14	No equivalent		mild aggression in stationary group	growl
15	No equivalent		strong aggression by silverback predator or other group	roar

Condensed and simplified from Marler (1976), after Van Lawick Goodall (1968) and Fossey (1972).

or curiosity. Roaring, given only by silverbacks, is an aggressive vocalization and the male often follows it up by a bluff charge. The hoot series, mentioned previously in connection with the chest-beating display, is again primarily a silverback display and occurs mainly during encounters between groups.

The observation that adult male gorillas vocalize more frequently than other group members correlates with the fact that silverbacks fulfil the major role in leading the group, protecting it against predators and in communications with other groups or lone males. Chimpanzees have a much more fluid social system, in which both visual and vocal displays are important for communication between sub-groups. Sub-groups frequently meet and merge or split, whereas the gorilla social group is a more compact entity. Adult and young chimpanzees of both sexes employ vocal communication in a variety of contexts. 'Pant-hooting', for instance, is a loud vocalization used mainly for communication between sub-groups. Marler and Hobbett have found that differences in pant-hoots produced by male and female chimpanzees are sufficient to allow a human observer to determine the sex of the vocalizer. Therefore, it is possible that a sub-group of chimpanzees can transmit information about its composition as well as its position by pant-hooting.

The information on gorillas summarized above refers exclusively to the mountain subspecies of the Virunga volcanoes. As no detailed studies have been made of the other two subspecies, it is not possible to say whether differences occur in the vocal repertoire of different gorilla populations to parallel the morphological and ecological divergences between them. Population differences in the vocal repertoire occur in other species, as for example in elephant seals studied by LeBoeuf, and white-crowned sparrows studied by Marler. White-crowned sparrows from differnt areas are recognizable by differences in their songs. 'Dialectic' differences might also occur between gorilla groups in different areas but this possibility remains to be investigated.

I have not described here communication during sexual behaviour or between a mother gorilla and her infant. It is important also to consider how the individual's communicatory abilities develop as it grows from infancy to adulthood. These topics fall within the scope of the next chapter.

Chapter 7

Reproduction and Infant Development

DURING the last twenty-five years gorilla births in captivity have gradually increased in frequency and a great deal of information has been gathered about the reproductive physiology and behaviour of these apes. The picture is still far from complete, however, and since it is so difficult to study reproductive processes in the wild, it seems likely that zoos and research centres will continue to play a vital role in this research. This chapter will review information on basic questions such as the age of onset of puberty in gorillas, attainment of sexual maturity, reproductive cycles, patterns of sexual behaviour, pregnancy, parturition, maternal care and infant development.

Onset of puberty and attainment of sexual maturity

Observations of captive female gorillas indicate that menarche (the first menstrual flow) occurs when they are between six and seven years old. Jorg Hess records that Goma, whose exact age is known, since she was born at Basle Zoo, first menstruated when she was six and a half years old. Two other females, Kati and Achilla, also reached menarche at between six and seven years of age. In contrast, there is one report by Charles Noback concerning 'Janet Penserosa', a female gorilla which did not reach menarche until about nine years of age. She had suffered from poliomyelitis, however, which had paralysed her legs, so it is possible that the onset of puberty had been delayed due to illness.

An examination of the data on age at first conception in eight captive gorillas indicates that females do not conceive until they are, on average, eight years and seven months old (Table 10). It seems that in gorillas, as in rhesus monkeys, chimpanzees and man there is a period of 'adolescent sterility' during which cycles occur, though they may be initially more irregular than in the mature female, and during which the female does not conceive. In chimpanzees, for instance, the period of adolescent sterility may last as little as four months, or as long as two years.

Table 10 also shows the age at which captive male gorillas first sire offspring; the average age is eight years, one and a half months in the six cases considered. These are approximate figures, however, and refer to a small number of animals. The exact age at which puberty begins in male gorillas has not been determined. Puberty in males is rather difficult to define, for the term refers not to some discrete event, but rather to a period of life during which a variety of hormonal, physical and behavioural changes occur. These changes often proceed

Table 10. Age at which the first fertile matings have occurred in captive gorillas

	Male	Female
Number of animals	6	8
Range	6 years– 9 years 6 months	6 years 9 months– 9 years 8 months
Mean	8 years 1½ months	8 years 7 months

These are only approximate values, estimated from various reports in the literature. Ages of animals which originate from the wild have doubtless been wrongly estimated by some authors. Hence it is very unlikely that a male gorilla could sire offspring when only six years old.

at different rates. Thus, in the gorilla, males grow rapidly and increase in weight during puberty, the reproductive system matures and there are secondary sexual changes, such as the gradual development of the silver saddle of hair on the back. The best way to pinpoint the onset of puberty is to discover when there is a pronounced increase in testosterone secretion by the male's testes. In male mammals this increase in hormone secretion typically precedes the completion of sperm production by the testes, just as the onset of ovarian cycles precedes attainment of fertility in the female. Unfortunately, no measures have been made of testosterone levels during development in male gorillas or orang-utans. In chimpanzees, however, McCormack has found that a large rise in testosterone occurs between six and nine years of age, followed by a further increase in fully mature males.

Problems of fertility in captive male gorillas

There have been reports that some adult male gorillas suffer from testicular atrophy in captivity. Four cases have been recorded by various authors, including Steiner, Koch, Antonius and Mckenny. The testes of the four gorillas examined were so degenerate in

Table 11. Details of six adult male great apes from which testes were removed at post mortem for histological study

Species and name	Date and place of death	Cause of death	Approximate age (years)	Weight (Kg)	Number of offspring	Condition of testes
Western lowland gorilla 'Oban'	Y.R.P.R.C. January 1976	Gastro-enteritis	12	79	0	Severely atrophic
Western lowland gorilla 'Jojo'	Chester Zoo March 1978	Pneumonia	18	N.D.	0	Severely atrophic
Western lowland gorilla 'Guy'	Z.S.L. June 1978	Anaesthesia – failed to recover after dental surgery	33	240	0	Partially atrophic
Common chimpanzee 'Freddie'	Z.S.L. October 1977	Euthanasia	16	N.D.	N.D.	Normal
Bornean orangutan 'Oscar'	J.W.P.T. July 1973	Ulcerative colitis	14	85	2	Normal
Bornean orangutan 'Boy'	Z.S.L. March 1976	Hepatic failure causes unknown	14	100	8	Normal

Modified from Dixson et al. (1980)

Y.R.P.R.C. = Yerkes Regional Primate Research Center
J.W.P.T. = Jersey Wildlife Preservation Trust
Z.S.L. = Zoological Society of London
N.D. = No data

For further details see text.

structure that the animals were undoubtedly sterile. Recently, at the Zoological Society of London, we have examined testes from three adult western lowland gorillas, one chimpanzee and two orang-utans which had died in captivity (Table 11). The testes from the orang-utans and chimpanzee were apparently normal whereas those from the gorillas exhibited varying degrees of atrophy (see Plate 19). In 'Guy', a silverback which had lived for over thirty years at the London Zoo, the testes were extremely small, measuring 27×18 mm (right) and 28×19 mm (left). The corresponding weights including the atrophic epididymes were only 6.85 grammes and 4.4 grammes respectively. The entire genital tract of this male was removed, yet it was not possible to identify with certainty either the prostate or seminal vesicles. Sections of the testes exhibited a diffuse nodular appearance, the seminiferous tubules were totally degenerate and fibrosed and none of the stages of spermatogenesis could be identified. Fine structural observations, using the electron microscope, revealed that the Leydig cells, which normally produce testosterone, were non-functional. 'Jojo', a much younger male which had died at

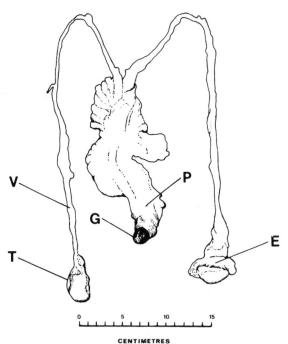

CENTIMETRES

Figure 44. Reproductive organs of an adult male western gorilla ('Guy') which died in captivity and which exhibited genital atrophy.

E epididymis; G glans penis; P shaft of the penis; T testis; V vas deferens. The seminal vesicles and prostate were atrophic and are not visible in this dissection. Author's drawing and specimen.

Chester Zoo, exhibited a similar testicular picture. In the third gorilla, 'Oban' from the Yerkes Regional Primate Research Center, the testes had a more normal appearance (see Plate 19). The seminiferous tubules were small and surrounded by large amounts of interstitial tissue, but this is not an indication of abnormality, since both Wislocki and Hall-Craggs have described these features in material from wild-shot gorillas. However, in 'Oban' sloughing and degeneration of the germinal epithelium was apparent and mature spermatozoa could not be identified in either the seminiferous tubules or epididymis.

The causes of testicular atrophy in captive gorillas are not known and a number of possibilities must be considered, such as dietary deficiencies, disease or psychological factors. It is possible, for instance, that keeping male gorillas in small enclosures or depriving them of social companions may adversely affect testicular function. Only three of the thirteen male gorillas in British collections are of proven breeding ability. Behavioural factors, or infertility in the female also affect breeding success, of course, not simply the male's fertility. I think it likely, however, that some non-breeding males are suffering from testicular atrophy. Much more information is needed on this subject. It is to be hoped that whenever a male gorilla dies in captivity, the reproductive organs will be preserved for examination.

The menstrual cycle of the gorilla

The menstrual cycle is a process common to the Old World monkeys, apes and man. The cycle lasts about four weeks, but age, health and social factors may affect its duration. In 1939, Charles Noback published the first account of the gorilla's menstrual cycle. He studied a young female whose cycles ranged in length from thirty-six to seventy-two days. Adolescent females often have more irregular cycles than adults, however, and much more detailed information on adult gorillas has now been obtained by Ronald Nadler of Yerkes Primate Center. He found that in eight western gorillas the cycle lasted an average of thirty-one or thirty-two days with a menstruation of one or two days.

Originally Noback had observed that the external genitalia of the female gorilla undergo slight but distinctive changes in swelling during her cycle, being most swollen at mid-cycle and least swollen during the menstrual phase (see Plate 20). Nadler found that the period of maximal labial swelling lasts for about two days at mid-cycle. These changes in the labia are important because they correlate with changes in hormone levels during the cycle and also with fluctuations in sexual behaviour.

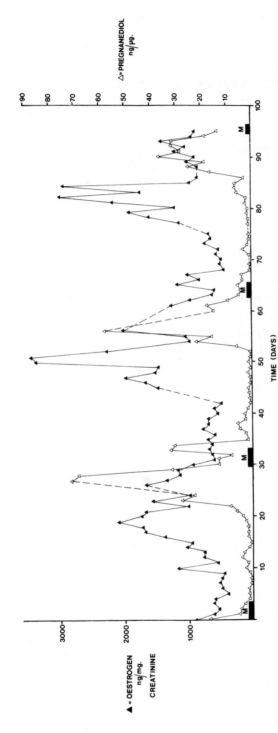

Figure 45. Cyclical changes in excretion of oestrogens and pregnanediol during three consecutive menstrual cycles in a female western lowland gorilla. Dark bars indicate menstruation (M). The hormonal data are approximate estimates of pregnanediol and of 'total oestrogens' in the female urine. The antiserum used to measure oestrogens cross reacts with oestrone, oestriol and oestradiol. Unpublished data reproduced by courtesy of R. D. Martin and R. C. Bonney.

Hormonal changes during the menstrual cycle of the gorilla have only been studied during the last few years. It is difficult to obtain blood samples regularly from captive gorillas and some workers have therefore measured hormones excreted in the female's urine. Preliminary results of one such study, by Robert Martin and Rosemary Bonney, are shown in Fig. 45. They measured concentrations of oestrogens and pregnanediol in daily urine samples from female western lowland gorillas. Pregnanediol, as Younglai, Collins and Graham have shown, is the major urinary metabolite of progesterone in the gorilla. Hence the graph gives a general idea of rhythmic changes in oestrogen and progesterone secretion by the female's ovaries during three consecutive menstrual cycles. Oestrogen levels are highest at mid cycle, which is when the female's labia are maximally swollen and she ovulates. Oestrogen levels decrease during the second (luteal) phase of the cycle, when the labia detumesce and the ovaries secrete increasing amounts of progesterone. More precise information has been obtained by measuring hormone concentrations in the blood of female gorillas. Nadler and his colleagues at the Yerkes Primate Center have recently completed such a study, and the reader is referred to their paper for a detailed treatment of this topic. It is evident that a preovulatory surge of oestradiol occurs in female gorillas, and a smaller peak during the luteal phase, just as in chimpanzees or man. In chimpanzees, however, there is a pink 'sexual skin' (see Plate 20) which is much more sensitive to circulating oestrogens than are the gorilla's labia. Hence increasing levels of oestrogen in the follicular phase cause water retention in the intercellular portions of the sexual skin so that, at mid-cycle, it may contain over a litre of fluid. Graham and his co-workers have found that ovulation occurs just as the female's sexual skin begins to deflate, and that deflation correlates with a decline in oestrogen levels coupled with an increase in circulating progesterone. In fact, it has been known for many years that progesterone antagonizes the effects of oestrogen upon the sexual skin. Cyclical changes in labial swelling are much less obvious in gorillas than in chimpanzees, it can be inferred, therefore, that they are controlled by similar hormonal mechanisms in both species.

Sexual behaviour during the menstrual cycle in gorillas

The mating behaviour of gorillas in captivity has been observed by Hess, Nadler, Thomas, Tijskens and others. Hess has made a valuable study of mating behaviour in a social group of gorillas, whilst Ronald Nadler has obtained a wealth of quantitative information by observing pairs of gorillas.

Jorg Hess studied a group of nine western lowland gorillas at Basle Zoo. This group consisted of two silverbacks, three adult females, a blackback and sub-adult male and an infant of each sex. He observed that adult females seemed to be sexually attracted to the dominant silverback and were most willing to mate with him during a three-day 'oestrus' in the middle of the menstrual cycle. In the period preceding oestrus, Hess noticed a change in the odour of the female and observed that the dominant male sniffed her vagina and armpits more frequently. The blackback and sub-adult males also touched the female's genitalia or armpits and then sniffed their fingers. It is possible that there is a change in the female's odour at mid-cycle which alerts males to her altered sexual status. In the 'pre-oestrus' phase, the female's behaviour also changed and she became visually attentive to the dominant silverback, posturing with her limbs held stiffly and staring at him. Hess noted that the female exhibited a distinctive facial expression with the lips pressed together and the mouth corners drawn in. This would seem to correspond to the lip-tucking or tight-lipped face which was described in the previous chapter and which occurs characteristically in situations of uncertainty.

The male also adopted a stiff-legged posture (that is, a strutting posture) and glanced repeatedly out of the corners of his eyes. The silverback occasionally displayed with a strutting run and struck at the female as he ran past. Sometimes this display was preceded by hooting and chest-beating. These behaviours have been described in Chapter 6 and it is clear that they occur in a variety of situations and not just in a sexual context.

During the 'oestrus' phase, Hess observed frequent incidences of 'inviting' and 'offering' by the adult female to the dominant silverback. One female employed a remarkable display in which she lay on her back, looking at the male and stretching one arm out towards him with the palm turned upwards. Occasionally she caught hold of the male and pulled him towards her. Hess also observed young gorillas using similar gestures as invitations to play, but their use as a sexual solicitation has not been described elsewhere and has not been observed in the wild. Perhaps this behaviour was an idiosyncracy of one particular female, a possibility which underscores the danger of basing generalizations about the behaviour of a species upon observations of a few individuals. The more typical method of inviting copulation, which has also been observed in wild gorillas, is for the female to crouch and back up towards the male so as to display her genitalia, and at the same time to look back at him.

More recently, Ronald Nadler of Yerkes Primate Center has

obtained extensive information on mating behaviour in captive gorillas. He observed nine female and four male western lowland gorillas during a three-year period. These animals were tested in pairs, a standard technique employed in work on sexual behaviour since it allows the experimenter to observe interactions in the simplest social setting without the added complications which occur in a group environment. The sexual status of the females was assessed by rating labial swelling on a scale of 0–2 (0= deflating, 1 = minimal swelling, and 2 = fully swollen labia).

Nadler observed that gorillas usually copulated in a dorso-ventral position, but that some pairs adopted a ventro-ventral posture (Fig. 46). He considered that this variability reflected the more complex brain development and superior cognitive abilities of gorillas, as compared with monkeys or lower mammals which employ stereotyped mating patterns. Copulations were usually brief, and males ejaculated in a single mount after performing about thirty-six pelvic thrusts. Both sexes vocalized during copulation and Nadler likened these sounds to the 'cooing of a dove'. He also noted that some gorillas showed marked sexual preferences for particular partners. This phenomenon has been reported in various primates such as rhesus and pigtailed monkeys and also in other mammals such as beagles.

Nadler's observations on cyclicity in gorilla mating behaviour confirm the general findings of Hess, Thomas, Tijskens, Reed, Gallagher and others, namely that matings are confined to a one to four-day period at mid-menstrual cycle. This, as we have seen, is the period of the cycle when the female's labia are maximally swollen. Some of Nadler's findings are shown in Fig. 47. The great majority of copulations were initiated by females and the frequency with which they 'presented' sexually to males was greatest when the labia were fully swollen and least when they were detumescing. These results are consistent with the idea that the female's willingness to initiate sexual interactions (that is, her proceptivity) and to accept the male's attempts to mate (that is, her receptivity) was greatest during this short period at mid-cycle. However, this interpretation is complicated by the fact that female gorillas may present to males throughout the cycle but that males accept a greater proportion of presentations when the female's labia are fully swollen. The female 'success ratio' (the percentage of presentations which elicit copulation) is therefore greatest at mid-cycle. This observation is consistent with the idea that female gorillas are more sexually attractive to males when the labia are swollen. The restriction of copulation to a brief phase at mid-cycle may depend on changes in the females' attractiveness and not solely upon changes in her own sexual behaviour.

Figure 46. Copulatory postures of western lowland gorillas. (*Top*) dorso-ventral; (*bottom*) ventro-ventral copulation. Author's drawings from photographs by R.D. Nadler.

The pronounced mid-cycle peak in the copulatory behaviour of gorillas is surprising. It has been suggested, by Beach originally and subsequently by many others, that in higher primates neural mechanisms which govern sexual behaviour are less influenced by gonadal hormones than in lower mammals. As we saw in Chapter 2, the cerebral cortex shows progressive enlargement in primates, and this trend finds its greatest expression in the apes and man. It would seem reasonable to suggest that the neocortex plays a greater role in controlling sexual behaviour and that the importance of gonadal

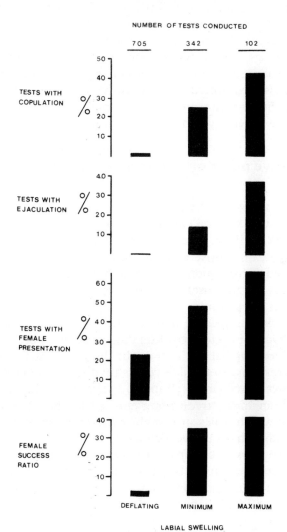

Figure 47. Changes in sexual interactions during the menstrual cycle of western lowland gorillas. For explanation see text. From data in Nadler (1975).

hormones in this respect has diminished during primate evolution. Many experimental studies are consistent with this idea, but those on the gorilla do not fit into such a simple scheme. Gorillas have large and complex brains, yet they confine their mating activities to a period of between one and four days in the middle of the menstrual cycle. One would have expected copulations to occur throughout the cycle as is the case in some macaque monkeys (for example, the rhesus and stumptail) or in humans. Clearly, the sexual behaviour of gorillas is greatly influenced by hormonal changes during the menstrual cycle but as yet we have little idea of the mechanisms involved. It is possible that the most important effects are not upon neural mechanisms in the female, but upon non-behavioural cues (for example, vaginal cues) which regulate her attractiveness to the male. Much research is needed in order to explain how hormones affect sexual attractiveness, proceptivity and receptivity in female gorillas. However, since these are rare and valuable animals, it will be very difficult to obtain the necessary experimental information.

Comparative notes on chimpanzee and orang-utan sexual behaviour

During the 1930s and 1940s the sexual behaviour of the common chimpanzee was studied extensively, using pair-testing techniques, and the work of Elder, Nissen, Yerkes, Young and Orbison is still the most complete on captive animals. More recently, however, Goodall has collected information on sexual behaviour in wild groups of chimpanzees, so that useful comparisons with the work on captive chimpanzees are possible.

The female chimpanzee may initiate copulation by presenting to the male, backing up towards him in a crouching posture with limbs flexed and looking back at him as she does so. The female may also present in response to a male's courtship display and Goodall lists six such displays by wild males; 'bipedal swaggering', the 'sitting hunch', 'glaring', 'branching', 'tree leaping', and 'beckoning'. Yerkes described a display in which males sit upright with thighs spread and the penis erect. Similar behaviour also occurs in the wild.

In captive pairs of chimpanzees, Yerkes found that either partner might initiate copulation but that in some cases males were very assertive and 'dominant' over females. In the wild, Goodall reports that 176 of 213 copulations were initiated by males. This is obviously very different from the situation in gorillas where males initiate copulations comparatively rarely. Male chimpanzees look at and sniff the female's genitalia, particularly when the sexual skin is swelling and, according to Goodall, 'in the period immediately following detumescence'. Chimpanzees typically copulate in a dorso-ventral position, though a number of variants of this pose are employed, and ejaculation occurs after about five to ten seconds. As the male nears ejaculation he utters panting sounds and the female may look back at him, making a squeaking vocalization and a 'grinning' facial expression.

Goodall has found that the copulatory behaviour of male chimpanzees is influenced by age and social factors. Thus older and more dominant males copulate more frequently with females which are at mid-menstrual cycle than adolescent males do. There is also some evidence that social factors and age may influence the frequency of mating behaviour in gorillas. In Hess's captive group, most copulations were performed by the 'dominant' silverback and females directed their sexual invitations towards this male. Schaller, however, observed only two copulations by wild mountain gorillas; both involved subordinate males and were not interfered with by dominant silverbacks. Recently, Harcourt and Stewart have published more extensive observations on wild mountain gorillas which indicate that mature males copulate more frequently with older females whereas blackbacked males mate more often with younger females which have not yet had their first offspring. The reasons for this are not clear, but it is possible that silverbacks inhibit younger males from mating with older females.

Both in the wild and in captivity the frequency of copulatory behaviour in chimpanzees exceeds that of the gorilla, and though some male and female chimpanzees may form consortships of short duration, generally they are 'promiscuous'. Mating may occur at any time during the female's cycle but significantly more sexual behaviour occurs during the follicular phase, when the sexual skin is enlarging, than takes place in the luteal phase, when the swelling detumesces. In some pairs of chimpanzees mating is much more frequent at mid-cycle when the sexual skin is fully swollen, but this is by no means the typical pattern for the species and matings may be frequent throughout the follicular phase. There is, therefore, no pronounced one to four day peak in copulations as occurs in the gorilla. Ovarian hormones doubtless play an important role in controlling the mating behaviour of chimpanzees, for Young and Orbison found that mating frequencies decreased markedly after they had ovariectomized three females. The mechanisms by which ovarian hormones exert their effects on chimpanzee sexual behaviour have not been studied.

Very little is known about the sexual behaviour of the pigmy chimpanzee. Savage, Bateman and colleagues have observed three pigmy chimpanzees in captivity and have recorded some aspects of sexual behaviour. Copulatory postures were variable and ventro-ventral copulations frequent, often accompanied by 'prolonged eye contact' between the participants. Although these observations indicate that the mating behaviour of pigmy chimpanzees differs from that of the larger species, it must be remembered that they refer to very few animals. Further, the only male observed was between six and a half and seven and a half years old and it seems unlikely that he was fully mature.

The sexual behaviour of orang-utans has not been so extensively studied as in gorillas or common chimpanzees. However, field studies by Mackinnon and Rijksen, as well as observations on captive orang-utans by Asano, Coffey, Heinricks and Dillingham, Jantschke and Nadler provide a general picture of mating behaviour in this species. Firstly, it seems that copulation may be a frequent and prolonged activity, at least in captivity, and that dorso-ventral and ventro-ventral postures are employed. Male orang-utans are very assertive sexually and more than one author has used the term 'rape' to describe their behaviour both in captivity and in the wild. The menstrual cycle has been studied by Collins, Graham and Preedy and it lasts, as expected, for about four weeks. In contrast to chimpanzees or gorillas, orang-utans do not show cyclical changes in sexual skin or labial tumescence during the menstrual cycle. Recently, Ronald Nadler has found that orang-utans mate throughout the female's menstrual cycle without pronounced cyclical changes in behaviour. However, he has also found that if the testing situation is arranged so that the female chooses whether she interacts with the male, then cyclical variations in her proceptivity are apparent.

In its wild state, the orang-utan's social organization differs considerably from that of the other two great apes. John Mackinnon observed in Borneo that adult male orang-utans lead solitary lives and associate with females only for sexual purposes. In Sumatra, however, males may stay with females during pregnancy, an observation which Mackinnon correlates with the fact that predators are more frequent in Sumatra and that male orang-utans may play a defensive role. Mackinnon also suggests that younger males range more widely than the massive fully mature males and mate more frequently. The older males, with fully developed cheek pads, are solitary and territorial and they employ a roaring vocalization to advertise their presence both to other males and to females.

Obviously, the three great ape genera differ tremendously from one another in their sexual behaviour, so that conclusions about one genus cannot be readily applied to another. This variability is not so surprising, however, when one remembers the great diversity of sexual behaviour exhibited by other primates. For instance, Bonnet monkeys, stumptails and rhesus monkeys all belong to the same genus (*Macaca*), yet their patterns of sexual behaviour are strikingly different, just as different as those of gorillas, chimpanzees and orang-utans.

Gestation in the gorilla

In order to determine the length of gestation in gorillas accurately it would be necessary to allow females to mate only once, at mid-cycle, so that the date of conception would be known. Such experiments have never been done, so we have only approximate estimates, which assume that the last day on which a pair mated was the date of conception. This method is inaccurate, because gorillas are known to copulate during pregnancy. Harcourt has observed such behaviour in wild mountain gorillas and Ronald Nadler observed one western lowland female, in captivity, which continued to mate until one week before giving birth. Nor do ovarian cycles cease immediately that a female gorilla becomes pregnant; Nadler recorded two cycles of labial tumescence, lasting twenty-nine and thirty days, following conception. This finding is not unexpected, for in 1943 Nissen and Yerkes reported genital swellings during early pregnancy in chimpanzees and pregnancy swellings have also been recorded in some Old World monkeys.

As can be seen in Table 12, estimates of the duration of pregnacy in gorillas range from 237 to 288 days. The points at the two ends of the range are atypical values, however, and if they are discarded, the remaining ten gestations range from 246 to 269.5 with a mean value of 257.6 days. This is probably quite an accurate estimate. Recently a female gorilla called Lomie gave birth at the London Zoo and her gestation period must have been between 254 and 257 days, since the exact dates on which she mated are known. The gestation length in chimpanzees is generally shorter than for gorillas. Nissen and Yerkes, after studying forty-nine cases, recorded a range of from 202 to 248 days and a mean of 226.8. Because the exact dates of conception are not known in these cases, the estimates may be inaccurate. No precise figures are available on the duration of pregnancy in orang-utans either, but estimates range from between 227 and 285 days. Orang-utans are unusual in that the sexual skin swells during pregnancy; a fact which was first recorded by Schultz in 1938. It is now known, as the result of researches by Lippert, that this swelling develops early in pregnancy, twenty-eight to forty days after the last menstrual

Table 12. Estimates of gestation lengths in the gorilla

Gestation length (days)	Source
237	Harding, Danford and Skeldon (1969)
246	Nadler (1975)
250	Nadler (1974)
252	Lang (1964)
254	Hopper, Tullner and Gray (1968)
255	Fontaine (1968)
258	Thomas (1958)
259	Mallinson, Coffey and Usher-Smith (1976)
266	Mallinson, Coffey and Usher-Smith (1976)
266.5	Mallinson, Coffey and Usher-Smith (1976)
269.5	Mallinson, Coffey and Usher-Smith (1976)
288	Lang (1959)

Mean 258.4

S.D. 13.06

Each example refers to one gestation. Where authors have given a range for the possible gestation length, an average value has been calculated.

S.D. = Standard deviation.

period. The swelling should therefore provide a useful means of diagnosing pregnancy in the orang-utan. Doctors often date gestation in women from the time of the last menstrual period. On this basis, Gibson and McKeown reach a figure of 280 days, but if one corrects this by fourteen days which is roughly when conception would have occurred, then a figure of 266 days is obtained.

A number of pregnancy tests have been devised which have as their basis the detection of chorionic gonadotrophin (HCG) in women's urine. Gorillas, like human beings, chimpanzees and baboons, excrete some type of chorionic gonadotrophin throughout pregnancy, and peak levels of hormone occur during the middle third of gestation. Pregnancy tests devised for human females are also effective for gorillas, but it has been reported by Martin, Seaton and Lusty that a test developed for use on sub-human primates proved more sensitive in determining early and late pregnancy in the gorilla than did the Pregnosticon Planotest, a technique developed for use on human urine. The latter workers also measured excretion of total oestrogens (oestradiol, oestrone and oestriol) in three pregnant female gorillas and found that hormone levels began to rise very early in pregnancy and continued to do so throughout the gestation period. This is an interesting observation, since it allows early detection of pregnancy and aids in determining the true date of conception. If similar studies are carried out on other apes we may obtain more accurate information on gestation length.

There are several physical changes which are useful as indicators of pregnancy in gorillas. Regular monitoring of body weight is valuable; thus Nadler has recorded an increase of 6.6 kilogrammes and Lang an increase of 7.5 kilogrammes in body weight of pregnant female gorillas. Towards the end of pregnancy the female's abdomen is noticeably enlarged and rounded in contour. About four months before she gives birth, the female's breasts may begin to enlarge and, in late pregnancy, some females squeeze milky fluid from their nipples.

Parturition

Gorillas in the wild and in captivity conceive and give birth at all times of the year. In 1970, Kirchshofer reviewed information regarding thirty-two births in captivity and found that twenty-two of them occurred during the 'warm' months of the year from April to September. I do not think, however, that there is sufficient information to say that gorillas have an annual birth peak. In fact none of the great apes are seasonal breeders; at least, field studies do not indicate that they are.

Parturition in the great apes is normally a rapid process. As can be seen from Table 13, gorillas, orang-utans and chimpanzees typically give birth in less than one hour and in most cases there is little sign that parturition is imminent. For this reason constant observation is necessary if births are to be observed and it is especially important that they should be, because captive gorillas do not always care for their offspring. One exceptionally long birth in Table 13 is the example recorded for a female gorilla by Mallinson, Coffey and Usher-Smith. In this case, labour lasted nine hours twenty-four minutes, but this was due to the fact that the position of the foetus in the birth canal was abnormal.

The rapidity with which the great apes give birth correlates with the fact that the head of the newborn is remarkably small in comparison to the dimensions of the female's pelvic canal (Fig. 48). In man, by contrast, labour may be prolonged and the baby's large head is often turned sideways to facilitate its passage through the canal.

There are several descriptions of parturition in the gorilla. The female first becomes restless, pacing back and forth or reclining, and mucus may be discharged from the vagina. Nadler reports seeing what appeared to be birth contractions, but these were not timed. The allantoic sac appeared first and the female touched and eventually burst this. She 'reclined on her forearms, knees, side or back' and although she quite often touched the vulval area during parturition

Table 13. Duration of labour in the great apes

Ape	Observation		Source
Gorilla	1	Keeper checked each hour, yet missed birth	Carmichael, Kraus and Reed (1962)
	2	No more than 30 minutes	Lang (1962)
	3	Within two hours	Rumbaugh (1967)
	4	Infant discovered 20 minutes after last check	Thomas (1958)
	5	9 minutes from when female first touched and licked vulva	
	6	9 hours and 24 minutes	Mallinson, Coffey and Usher-Smith (1976)
	7	Keeper checked one hour previous to birth but saw no sign of labour	
	8	2 hours 53 minutes	Nadler (1974)
	9	'No obvious signs of labour'	Nadler (1975)
Chimpanzee	1–47	40 minutes to 8 hours	Nissen and Yerkes (1943)
	48	40 minutes	Lemmon (1968)
	49	4 hours 45 minutes	Wyatt and Vevers (1935)
	50	1 hour 20 minutes	Elder and Yerkes (1936)
	51	About 30 minutes	Montane (in Elder and Yerkes 1936)
Orang-utan	1	From 25 to 35 minutes	Graham-Jones and Hill (1962)
	2 & 3	Three to four hours for first twin, about one hour for the second	Heinrichs and Dillingham (1970)
	4	'No sign that [birth] was imminent'	Chaffee (1967)
	5	'Gave birth very easily'	Asano (1967)

Modified from Lindburg and Hazell (1972) with additional information on gorillas.

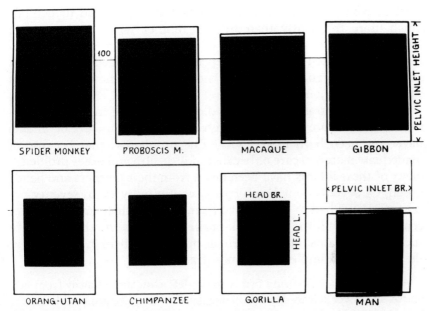

Figure 48. Diagram showing the relationship between the diameter of the pelvic inlet of various adult female primates and the length and breadth of the newborn's head. All diagrams have been reduced to the same pelvic inlet breadth. After Schultz (1969).

she did not use the hands to assist delivery of the infant. Similar but less detailed accounts have appeared concerning parturition in other captive specimens, and the general picture has been confirmed by Kelly Stewart's observations of parturition in a wild mountain gorilla.

Maternal care in the gorilla

Female gorillas lick the infant immediately after giving birth in order to clean away the birth fluids. All the great apes show this behaviour, and indeed it is a common mammalian pattern, so that Mantagu's statement to the effect that the great apes resemble man in not licking the newborn is incorrect. Gorilla mothers have also been observed to eat the placenta following the birth.

Hess records that, in captivity, female western lowland gorillas pay particular attention to the genitals of their infants, and that touching and mouthing of this area is common until the offspring is about three years old. How typical this is of wild gorillas we do not know for Schaller rarely observed such behaviour during his field work on mountain gorillas. Hess noted that genital manipulation by the mother functioned to keep the infant clean, to stimulate urination and, when performed roughly, as a means of punishment. One of the gorillas, whose infant was male, touched and inspected its genitals much more frequently than another mother which had a female offspring. This is an interesting observation, since it implies that infants are treated differently depending upon their sex, but not enough mothers and infants have been studied to know whether it is a general phenomenon.

It may seem strange that detailed observations of maternal behaviour in gorillas are scarce, although quite large numbers of infants have now been born in captivity. The problem is that because of inadequate maternal care or because of ill-health and other problems, many of these infants have been taken from their mothers and hand-raised. Of seventy-four live births in captivity by 1974, seventy per cent of infants had to be removed from their mothers and eighty per cent taken within the first week of life. Often the mothers failed to suckle their infants, ignored them or mistreated them in some way as the following examples will demonstrate:

Achilla held her offspring firmly in her left arm, turned away from her, and as she never released her vice-like grip it was quite impossible for the baby to reach the nipple. On the following day, when the infant expressed its hunger by whimpering and shrieking repeatedly, she held its little mouth shut with her other hand.

Ernst Lang

Though Vila kept the infant with her, she gave little indication that she knew what to do with it. During the first 20 minutes following birth, for instance, Vila repeatedly and with great persistence pulled the infant about by a handhold on the umbilical cord.

Duane Rumbaugh

During the first nineteen hours of life the infant was observed to attempt to suckle on twenty-two separate occasions, but apart from two occasions when he managed to gain the nipple before being pulled away, in the majority of cases the mother held him far too high.

Mallinson et al.

Why do these problems occur in captivity? Part of the answer may lie in the fact that most adult female gorillas in captivity were originally captured as infants or juveniles. Some have grown up with no opportunity to observe older females care for their babies. Some have had little contact of any kind with other gorillas. Rhesus monkeys and chimpanzees which have been reared in social isolation are sexually and maternally inept as adults. I am not suggesting that female gorillas in zoos have been reared in such severe isolation as the rhesus monkeys studied by Harlow, but some of them have grown up in partial isolation and lack social experience. A second possibility to be considered is that the confined conditions of many zoo enclosures and the boredom experienced by gorilla mothers in some way disturbs their maternal behaviour. Recently a group of gorillas has been established in a large open-air compound at Yerkes Field station. The females in this group show a high degree of competence in caring for their offspring, perhaps because their social environment is better than that experienced by most gorillas in zoos.

When Ronald Nadler reviewed all the information on maternal behaviour in captive gorillas he reached the conclusion that early social isolation had affected their abilities as mothers, but that these effects were less marked in females which had been reared with other young gorillas. Females with their first offspring (primiparous females) did not, however, require the experience of seeing other mothers care for their babies in order to develop adequate patterns of maternal care themselves. Colo and Goma were the first two gorillas born in captivity and both were separated from their mothers and reared with other youngsters. When they reached adulthood both these females had babies and showed good maternal responses, though Colo's infant was removed because of fears that it would contract tuberculosis. In July 1976 Lomie, a female gorilla at London Zoo, had her first infant and she displayed a high degree of competence in caring for the baby. She licked the infant clean following the birth, devoured the placenta and held the infant correctly so that it could reach her breast. After three weeks there were signs, however, that Lomie was neglecting the baby and so it was removed (see Plate 21).

Though captive primiparous females can make competent mothers it does seem that multiparous females are generally more proficient in this regard (Table 14). Multiparous females abuse their infants less, suckle them more readily and so are less likely to have their infants taken away. Primiparous females produce a higher percentage of premature or stillborn offspring and those infants which are removed in the first week are taken because mothers do not look after them properly. When infants are taken from multiparous mothers, it is usually because of infections or fear of illness. The behavioural differences between primiparous and multiparous females are mainly due to experiential factors, since, when females have been studied with successive infants, it has usually been found that they gradually improve in ability to care for the offspring. Plate 22 shows N'Pongo, a female western lowland gorilla kept at the Jersey Wildlife Preservation Trust. The first three offspring produced by this female had to be removed and hand-reared but she has succeeded in raising the fourth baby successfully.

Table 14. Percentage of captive primiparous and multiparous gorilla mothers exhibiting various degrees of maternal competency

	Primiparous females (n = 30)	Multiparous females (n = 34)
Abuse	20 (n = 6)	6 (n = 2)
Neglect – inadequacy	20 (n = 6)	21 (n = 7)
Clean – support infant	80 (n = 24)	82 (n = 28)
Ventro–ventral contact	60 (n = 18)	76 (n = 26)
Suckle infant	40 (n = 12)	68 (n = 23)
Retain infant at least one week	33 (n = 10)	59 (n = 20)

After Nadler (1975).

It is not known whether any of these findings are applicable to groups of gorillas in the wild, but it is important to consider this possibility. In the last chapter we saw that mortality rates are high among gorillas during infancy. This could result, in part, from a lack of maternal expertise among primiparous females. It is also important to remember that gorillas are slow breeders. If a female is successful in rearing her infant then the interbirth interval is approximately four years. The female is able to conceive again more rapidly, however, if the infant perishes or is removed. Reduction in interbirth intervals was noted in N'Pongo at Jersey Zoo and Lomie at the London Zoo, probably because their infants were removed soon after birth and hand-reared. If we assume that a female gorilla lives about thirty-five years then she might rear seven infants successfully during this time. It is probable that female gorillas remain active reproductively for most, if not all, their lives. The same is apparently true of the chimpanzee and orang-utan. There is one record of a female chimpanzee at London Zoo which gave birth when she was approximately twenty-nine years of age. As far as I know, the equivalent of the human menopause has not been recorded for any of the great apes.

Behavioural development in infant gorillas

The newborn gorilla is apparently helpless and its limb movements are unco-ordinated. The eyes show co-ordinated movements and a weak pupillary reflex, but the infant seems unable to focus its gaze correctly or to follow moving objects. It displays a startle reponse to loud noises, but cannot localize the source of the sounds.

Nevertheless, infant gorillas do possess a number of behavioural reflexes which are of great importance for survival. Stephen Carter studied these reflexes in two baby male gorillas which were born at the Jersey Wildlife Preservation Trust and were taken from their mothers very soon after birth. His observations are particularly illuminating because, as a paediatrician, he was able to make direct comparisons between the behaviour of infant gorillas and that of human babies.

Baby gorillas show rhythmic movements of the head in search for the mother's breast, the 'rooting' reflex which occurs in many mammals. As it searches for the breast the offspring may make soft vocalizations and the mother may respond to these by lifting and positioning the baby correctly so that it can suckle. A suckling reflex is also present and indeed in the case of the human foetus, this response has been elicited at three months by stimulating the inside of the lips or the tongue.

In contrast to reports by Lang that the newborn gorillas have weak grasping reflexes, Carter and others have observed that the grasp and suspension reflexes are strongly developed. Baby gorillas can grasp objects with either the hands or feet and hang suspended with no other means of support (see Plate 23 and Fig. 49). Nadler observed that one newborn supported her weight by one hand for three minutes, fifty-two seconds and could hang by one foot for eight seconds. By about three weeks of age the foot grasp has increased in strength to equal that of the infant's hand. Human neonates also display hand grasp and suspension reflexes but in a much weaker form. At three months, the human foetus exhibits such a grasp reflex, and babies born prematurely have a stronger grasp reflex than do full-term neonates. The grasp and suspension reflexes are already waning when a human baby is born and Carter states that they disappear completely after about three weeks. In gorillas, however, these reflexes persist until two or three months of age and then gradually merge into voluntary movements.

The adaptive nature of these responses is clear enough when one considers Battell's statement, no less valid today than in 1625, that 'the young pongoe hangeth on its mother's bellie'. The baby gorilla, like many other baby primates, must cling tightly to its mother, since she may run or climb occasionally and cannot always offer the infant full support. Mortality rates are higher during infancy than at any other phase of the gorilla's life, so mechanisms which assist infant

Figure 49. Foot-grasp reflex in an infant male western gorilla. Drawn by J. Ito from Carter (1973).

survival have high selection value. The occurrence of grasping reflexes in human infants may be viewed as an evolutionary remnant, a reminder of times when man's ancestors were more hirsute and when infants clung tightly to their mothers. Though the human baby lacks an opposable big toe and obviously cannot grasp objects with its feet like a gorilla, a remnant of the grasp reflex can be elicited if the observer's thumb is pressed against the sole of the baby's foot, as shown in Fig. 50. The reflex no longer has any practical function, though paediatricians find it useful as an indicator of infant development.

A 'startle' or Moro reflex is present in newborn human beings and gorillas, but it occurs more rapidly in the latter. If a baby gorilla is held belly uppermost and its head allowed to drop backwards so as to give an illusion of falling, then the baby extends its arms, spreads its fingers and makes an embracing movement so that the hands are brought together. The infant also raises its legs and attempts to grip with its feet. The response clearly functions to enable the baby to grasp something for support when it feels in danger of falling.

Figure 50. Foot-grasp reflex in a newborn human. Drawn by J. Ito from Carter (1973).

Baby gorillas have much greater strength in their back and neck muscles than do human neonates, as Carter found when he tested infants by ventral suspension; holding them in a prone position by placing one hand under the abdomen. When the baby gorillas were suspended in this way they were able to arch their backs and raise their heads above the horizontal plane, whereas in human newborns the head hung down slightly and the back appeared a little rounded. Not until about eight weeks of age could a human baby equal the performance of a newborn gorilla.

Infant gorillas undergo rapid behavioural development in their first year as can be seen from Table 15. Thus the infant's gaze is co-ordinated by three to six weeks and co-ordinated reaching movements may be seen by seven to eleven weeks of age. By seven to fifteen weeks the infant can sit up and by fourteen to twenty-five weeks it is able to walk quadrupedally. A knuckle-walking posture is employed, as in adult gorillas, and not a flat-handed posture such as is seen in monkeys or human children during quadrupedal locomotion. I stress that the estimates shown in Table 15 can only be approximate for they refer to a small number of infants and, moreover, mostly to captive, hand-raised individuals.

Infants raised with their mothers at first cling or are carried in a ventro-ventral position (Plate 22) but later they begin to ride on the mother's back. The first solid food is taken at about two and a half months of age. At around six months the infant starts to climb on its own and leaves the mother for short periods. In the last chapter we

saw that the infant mountain gorilla may make its first attempt at
nest building at eight months of age but it does not sleep in its own
nest separate from its mother until it is between two and three years
of age.

Many patterns of social behaviour mentioned already with reference to the behaviour of adult gorillas make their appearance during
the first year of life. Laughing and smiling appear during the first
two to five weeks after birth and once the infant starts to walk
it begins to participate in play, chasing, wrestling and mock-biting
with other youngsters. Redshaw has noted the typical 'play face' at
as early as fourteen weeks in the gorilla. Schaller saw a number of
kinds of solitary and social play in juvenile mountain gorillas including the 'snake dance' during which animals walked in a line, each
with its hands on the back of the one in front.

Chest-beating, or drumming with the hands upon objects begins
during the first year of life. It has been observed at between twenty-
six and forty-five weeks and the strutting walk with lip-tucking has
been recorded at thirty-four weeks in captivity. These behaviours are
seen when infants are playing or interacting socially in some way.
The patterns gradually become more perfect, partly one presumes as
a result of practice and partly as a result of physical development.
The typical adult 'threat face' with mouth open and teeth exposed
has been observed by forty-three weeks and if the infant is thwarted,
'temper tantrums' apparently occur from about twelve weeks of age,
but are not so common as in infant chimpanzees.

Sexual patterns also appear quite early in development. Hess, for
instance, saw 'play copulations' by a female aged one and a half years,
and Schaller reports mounting and thrusting by a two-and-a-half-
year-old male with a seven-month female. These observations are
not unexpected because it is known that neural mechanisms which
control sexual behaviour are organized during foetal life in primates
such as macaques. Androgens produced by the foetal testes during a
'critical period' influence the organization of pathways which mediate
patterns of masculine sexual behaviour. Indeed, male macaques and
baboons less than one year of age have been observed to mount adult
females, perform pelvic thrusts with the penis erect and achieve
intromission. The patterns are not expressed with full intensity and
frequency, however, until the testes become fully active at puberty.

The behavioural development of infant gorillas summarized so far
contrasts with that of human infants in being much more rapid and
the same is generally true of chimpanzee and orang-utan infants. A
review of comparative work on behavioural development in the great
apes and man is beyond the scope of this book. I wish to select only

Table 15. Summary of information on behavioural development of gorillas during their first year

Behaviour	Time of first appearance (range in weeks)
Laughs and smiles	2–5
Localizes sound source	3–10
Follows moving objects with gaze	3–6
Gazes at own hands	3–6
Co-ordinated reaching	7–11
Eye-hand-mouth co-ordination	9–10
First attempts at creeping	4–8
Sits upright	7–15
Quadrupedal creeping	8–17
Quadrupedal walking	14–25
Stands bipedally	14–23
Chest-beating	26–45
Strutting and lip-tucking	34–41

The ranges are approximate and refer to small numbers of infants, most of which were born in captivity and hand-raised.

Sources of information: Frueh (1968), Hinde (1971), Hughes and Redshaw (1973), Kirchshofer et al. (1967, 1968), Knobloch and Pasamanick (1959), Lang (1962), Redshaw (1978) and Usher-Smith, King, Pook and Redshaw (1976).

one example, an elegant study recently carried out by Margaret Redshaw, of cognitive development during the first eighteen months in gorillas and man.

Redshaw observed four gorillas born at Jersey Zoo and two human babies. She tested them at monthly intervals using four graded scales derived from the work of Piaget on children. Scale 1, 'visual pursuit and object permanence' consisted of fourteen steps ranging from visual tracking of moving objects up to complete tasks in which toys had to be located after they were moved from place to place hidden in a container. Gorillas and humans passed through the same stages in developing the concept of 'object permanence', but the apes embarked upon the developmental process at an earlier age. Visual tracking of moving objects, for instance, was first observed at six weeks in gorillas but not until ten weeks in man. In step six, infants were set the task of finding a toy after watching the experimenter place it under one of two cloths. The gorillas achieved this at an average age of twenty-four and a half weeks in contrast to thirty-two weeks in human infants. The most difficult step, in which the subjects had to locate a toy after it had been moved from place to place concealed inside a container, was accomplished at forty-three and a half week by gorillas but not until fifty-four weeks by humans.

Results obtained from testing gorilla and human infants on the other three developmental scales confirmed the finding that both species pass through similar stages of early behavioural development and that these stages usually occur in the same order. Gorillas, however, began to develop earlier than the human infants on all four scales measured. Human infants lag behind gorillas in the development of nervous control of movement. By seventeen and a half weeks gorillas are able to use locomotion to reach a toy but this capacity does not emerge until about forty weeks in man. Redshaw found, however, that human infants were superior in certain aspects of intellectual development. The two children in her study could use a rake to draw in objects placed beyond their reach when they were fifty-four and fifty-eight weeks old. Three of the four gorillas never mastered this problem, however, and the fourth did not do so until 104 weeks of age. As discussed in Chapter 5, the poor performance of gorillas in tool-using problems is partly attributable to their inferior manipulative abilities or lack of motivation and not solely to an intellectual incapacity to comprehend the solutions of such tasks. Gorilla infants were ahead of humans in developing the capacity to make detours in order to obtain an objective (the 'round-about methods' behaviour discussed in Chapter 5) but fell far behind the latter in developing the ability to build towers of blocks or to take account of gravity by allowing a wheeled toy to roll down an inclined plane. Indeed none of the apes built a tower of blocks and only one male succeeded in using the ramp and toy when he was fifty-four weeks old. Whereas human infants learned to interact with the experimenter, for instance by handing her back a mechanical toy in order to watch it demonstrated again, the gorillas never exhibited such behaviour. Nor do young gorillas indulge in reciprocal play involving objects when they are playing together or with their mothers; a major difference from human children. After the first year the intellectual gap between the two species widens rapidly, particularly, of course, when human infants begin to talk.

Chapter 8

Conservation or Extinction?

THE *Red Data Book* lists the gorilla as a *vulnerable* species, a term which covers a variety of possibilities. It includes 'taxa of which most or all of the population are *decreasing* because of over exploitation, extensive destruction of habitat or other environmental disturbances' or which are 'still abundant but *under threat*'. The mountain subspecies is placed in the category of *endangered* taxa 'whose numbers have been so drastically reduced that they are deemed to be in *immediate danger of extinction*'.

Hunting and habitat destruction have been the major causes of the gorilla's decline. In the past, gorillas were hunted to obtain trophies, or because skins and skeletal material were required by museums. Carl Akeley, for instance, was a taxidermist who between 1896 and 1926 went on five collecting expeditions to Africa and who prepared exhibits for the Field Museum of Natural History in Chicago and for New York's American Museum of Natural History. A hunter himself, Akeley none the less reached the conclusion that big-game hunting constituted a major threat to many of Africa's large mammals. Of gorilla-hunting he was to comment: 'killing gorillas cannot possibly be considered sport: the animals are easily located with the help of native guides in the regions they inhabit, easily approached and easily killed.' His criticisms fell on deaf ears amongst the hunting fraternity and, naturally enough, were opposed by those who profited from organizing big-game safaris. Fortunately the Belgian government, which then administered the Congo and Rwanda-Urundi, was most sympathetic and in March 1925, King Albert created Africa's first National Park. This was also the first step taken to protect gorillas, for within the Albert National Park lay the Virunga volcanoes, home of the mountain gorilla.

Big-game hunting or collecting for scientific purposes is no longer a major threat to the gorilla. Gorillas are hunted in some areas, either for meat or because they occasionally raid plantations. However, the two enduring problems are undoubtedly the destruction of the gorilla's natural habitat, without adequate provision of protected areas and, to a much lesser degree, the capture of living specimens. An unrestricted traffic in live gorillas would cause their extinction, because the methods used to capture them involve killing adults in order to secure the infants. For instance, Schaller records that: 'In about 1948, officials organized the killing of some sixty mountain gorillas near Angumu to obtain eleven infants, only one of which survived.' It is sad to reflect that mountain gorillas had been afforded complete protection by international agreement in 1933, for even then they were considered to be on the verge of extinction. It is clear that the trade in wild gorillas perpetuates their slaughter and that protection laws are useless unless actually enforced. As Caldwell commented in 1934: 'International conferences for the protection of the fauna of Africa are not new. The first one was held in London in 1900, but since none of the powers could ratify the decisions, it did not achieve much.' In many areas of the gorilla's distribution range expanding human populations are clearing the forest for agriculture, timber extraction or industrial development. To have any hope of success, conservation programmes should be based upon field surveys, followed by more detailed studies of the gorilla's ecology and behaviour. Unfortunately, as should be apparent from earlier chapters, field studies of the gorilla have been largely restricted to the mountain subspecies. The following sections will summarize what is known about the status of all three gorilla subspecies. In many cases each population must be dealt with separately, because of the discontinuous distribution of gorillas and because, until recently, they occurred in ten countries with differing views on the value of conservation. Since complete information is only available on one subspecies, the mountain gorilla, I shall deal with this first.

The mountain gorilla (*Gorilla g. beringei*)

Mountain gorillas are found in two areas; to the west of Lake Kivu in the Kahuzi–Biega reserve (Fig. 51) and in the nearby Virunga volcanoes (Fig. 52). The Kahuzi–Biega reserve was made a National Park in 1970 and covers an area of 70,000 hectares. It contains approximately 200 gorillas. Long before the area became a park the Conservateur, Adrian Deschryvver, paid guards out of his own pocket to protect the diminishing gorilla population. Deschryvver and his trained guards regularly take parties of tourists into the Kahuzi–Biega park to watch gorillas. Various groups have been habituated and visitors are assured of a sight of the animals. Given the success of this

venture and the efficiency with which Zaire administers the National
Park it seems that this small population of gorillas is secure for the
moment. Mackinnon, who visited the area in 1975 has commented
that the park is too small for the long-term maintenance of a viable
gorilla population. He also points out the danger of inbreeding in
small isolated populations of gorillas such as those at Kahuzi or in
the Virunga volcanoes.

The eight Virunga volcanoes form three separate groups, all of
which lie within areas of National Park or Reserve status. To the west

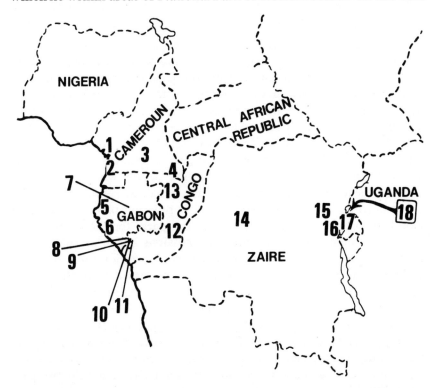

Figure 51. Rain forested National Parks and Reserves in countries where gorillas occur.
Cameroun 1 Douala Edea Reserve 2 Campo Reserve 3 Dja Reserve 4 Mouloundou
Reserve
Gabon 5 Wonga Wongue National Park 6 Petit Loango Reserve 7 Okanda National Park 8
Mt Fouari Reserve 9 Nyanga Reserve *Congo* (Brazzaville) 10 Mt Fouari Reserve (straddles
border with Gabon) 11 Nyanga Reserve (straddles border with Gabon) 12 Lefini Reserve
13 Ozala National Park *Zaire* 14 Salonga National Park 15 Maiko National Park 16
Kahuzi-Biega National Park 17 Virunga National Park
Rwanda 17 Parc National des Volcans
Uganda 17 Ugandan Gorilla Sanctuary. 18 Impenetrable Central Forest Reserve.

This list is not exhaustive. Inclusion of a Reserve or National Park on the map does not
necessarily mean that it contains or affords protection to gorillas. For further details see
text.

are the two youngest and still active cones of Nyirangongo and Nyamulagira. Gorillas are not found on their slopes, although tourists occasionally visit the area to see Nyirangongo's larva lake. The heartland of the Virungas is formed by three mountains; Mikeno, Karisimbi and Visoke (see Plate 24). The area is divided politically by the Rwanda–Zaire border which extends from the summit of Visoke to that of Karisimbi. The eastern group of volcanoes, Sabinio, Gahinga and Muhavura, lies along the boundary between Rwanda and Uganda. The borders of both countries meet that of Zaire at the summit of Mt Sabinio.

It is unfortunate, as far as conservation efforts are concerned, that the Virungas are divided between three countries. There is the Virunga Park in Zaire (formerly the Albert National Park), the Parc National des Volcans of Rwanda and the Ugandan Gorilla sanctuary, which was established in 1930. Added to the problems caused by a divided administration, the reserves lie on fertile soil in an area which is over-populated. In Rwanda, for instance, there is now a population density of 600 per square mile and the country will soon contain five million people. The formation of the gorilla reserves was inevitably an act foisted upon an uncomprehending people when there was a pressing need for land for agriculture, grazing, rights-of-way and wood-cutting. There was also a small population of Batwa pygmies who lived as nomadic hunters inside the park, for the forests were their traditional home. Territorial concessions began in 1929, when 10,000 hectares of reserve were sacrificed for agriculture. In 1950, the Ugandan portion of the reserve was decreased from thirty-four to twenty-three square kilometres when the boundary was raised from the 2,440 to the 2,740 metre contour line. Much of the habitat suitable for gorillas was lost. In 1969, 10,000 hectares was appropriated from the Rwandan Park in order to grow pyrethrum. According to Harrison: 'Ten thousand families were settled and 100 kilometres of roads were built during 1969/70 within the park boundaries.'

In 1980, Harcourt reported that the Ministry of Agriculture in Rwanda had plans to appropriate up to forty per cent of the remaining Parc des Volcans. He also pointed out that even if the entire area is taken for cultivation 'it would supply land for only one quarter of one year's increase in the population.'

Considerable efforts have been made to census the gorilla population of the Virunga volcanoes. In 1970, a project was initiated by National Geographic, aided by grants from the World Wildlife Fund and the Fauna Preservation Society. Dr Dian Fossey supervised this work. I took part in the final phase of the census in 1973, the aim being to map the distribution of gorillas in the Virungas and assess their chances of survival. This necessitated a sector search of the entire area from a series of tented camps. All gorilla sign was noted and fresh trails followed until the animals were located. Since gorilla groups range over large areas and ranges overlap extensively, it was vital to identify each group by making 'noseprints' (line drawings of the nasal area) of as many individuals as possible. Consistent and

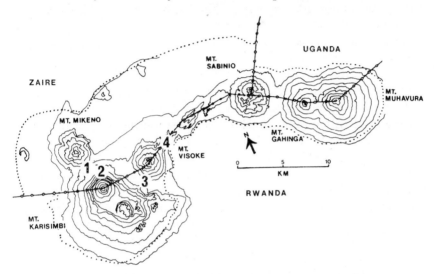

Figure 52. Map of the Virunga volcanoes. The present Park boundaries are indicated by a dotted line. Huts have been built in several areas. 1 Kabara, 2 Rukumi, 3 Karisoke Research Centre, 4 Ngezi.

consecutive counts of night nests provided the best information on group sizes.

A total of 174 gorillas was found elsewhere than on Mt Visoke, for which Dr Fossey had recorded a count of ninety-six animals. This gave a total of 270 for the minimum population of gorillas in the Virungas. Only thirteen animals were found in Uganda, seventy-four in Rwanda and 183 in Zaire. These figures are not fixed, of course; they will vary as the animals move back and forth across the borders. However, they provide a good picture of how the gorilla population was distributed in 1973. Because the numbers of gorillas on Sabinio and Muhavura are so small, and human disturbance is high in these areas (Table 16), their future seems bleak. On Mount Mikeno the largest population of gorillas appeared to be well protected in 1973 although recent reports indicate that the situation has deteriorated. On Visoke, the population has survived principally due to the efforts of Dr Fossey and workers of the Karisoke Research Centre, established in 1969. Her scientific articles and those by Dr Harcourt, as well as films made by the National Geographic Society have done a great deal to promote the cause of gorilla conservation.

By the end of 1973, about eighty-five per cent of the Virunga parks had been searched and, theoretically, the count of 270 gorillas might be fifteen per cent short of the actual total. However, the unsearched areas were to the west of Karasimbi in Rwanda and north of the Visoke-Sabinio saddle in Zaire. The former area was known to be over-run with cattle and poachers, while the latter area consists mainly of bamboo and 'dry colonizing forest'. It is unlikely that many gorillas survive in these unfavourable regions, although it is possible that there are some in the dry colonizing forests, since the 1973 census showed that gorillas sometimes use this habitat.

Illegal use of the Virunga Parks has always been a major problem. In the past this mainly involved damage caused by cattle grazing inside the park. Poachers also operate in the parks and signs of their activities, such as wire noose traps, are frequently to be seen in some areas (Table 16). The destruction of suitable habitat by cattle and human disturbance inside the parks has contributed to the decline of the gorilla population. In recent years a number of gorillas have also been killed to obtain their heads and hands for sale to tourists or European residents. These gruesome incidents prompted the Fauna Preservation Society to establish a special 'Mountain Gorilla Project' in 1978. In association with several other organizations, FPS has raised over £100,000 for use in efforts to conserve the mountain gorilla. The money is being used in several ways, for instance for an education programme since, as Harcourt points out, 'many Rwandans do not know what a gorilla is, let alone that it lives in their country and is endangered.'

Efforts are also being made to strengthen the management system of the Parc des Volcans. Guides are being trained and several groups of gorillas have been habituated to the presence of tourists. This is an extremely important venture, since it will be essential to develop a tourist industry if the Virunga parks are to survive. In the past, little effort has been made to develop tourism. Few are aware of the spectacular scenery of the extinct volcanoes, of the larva lake in the active crater of Nyirangongo, or of the fact that elephant, buffalo and other big-game besides the gorilla can still be found in the Virungas. It would obviously assist matters if the three countries which administer the Virunga parks could join forces to protect and develop them. Harcourt reports that gorillas are still being killed and infants captured for sale in the Zaire section of the park and that the gorilla population may now have fallen to around 200.

Table 16. Examples of illegal use of the Virunga Parks

AREAS STUDIED	Saddle between Mt Gahinga and Mt Muhavura	Saddle between Mt Gahinga and Mt Sabinio	South of Mt Sabinio	West of Mt Sabinio	South of Mt Karisimbi
Man-hours worked at each site	70	40	12	20	16
No. of poachers	6	7	–	–	–
No. of smugglers	42	169	–	–	–
Traps found	18	22	27	13	30
Cattle herds	5	7	3	3	4
Total no. of cattle	134	209	70	26	56
Wood/grass cutters	11	47	4	–	9
Herdsmans' shelters	16	42	5	3	7

Data: collected in 1972. From Groom (1973).

The eastern lowland gorilla (*Gorilla g. graueri*)

There has been no full survey of either of the two lowland subspecies and so estimates of their numbers and distribution can only be approximate. In 1959 Emlen and Schaller proposed that the population density of eastern gorillas was about one per square mile. In 8,000 square miles of forest there might, therefore, be about 8,000 gorillas. With scholarly caution, they suggested a range of from 5,000 to 15,000 animals, which included the Virunga population. Sadly, the correct estimate may be at the lower end of the range. If similar methods were applied to the Virungas, using Dr Fossey's count of ninety-six gorillas on Mt Visoke as a basis, then the estimate for the whole area would be 1,000 animals. One might then use caution and suggest a minimum count of 600, but in fact this is more than twice the actual number recorded by census expeditions.

The eastern gorilla is much more numerous than the mountain subspecies and is widely scattered across vast, thinly populated areas of forest. Although there is little pressure on its range, gorillas do seem to congregate in areas of secondary forest created by man's shifting cultivation. Jacques Verschuren published an article in 1975 in which part of the range of the eastern lowland gorilla is shown as a hatched circle west of Lake Kivu, surrounded by outwardly radiating arrows. He thus indicated his uncertainty about the distribution of gorillas in the area. A more definite distribution map is shown in Fig. 10 (page 27). Until field surveys are conducted, the status of the eastern lowland gorilla will remain uncertain, so I shall concentrate on those areas where efforts have been made to conserve these animals.

Zaire has set aside 1,000,000 hectares in the eastern part of the country as the Maiko National Park (Fig. 51). Gorillas have a wide distribution in the lowlands of eastern Zaire, from the Ulindi river in the south and extending northwards almost to the equator. The Maiko National Park does contain gorillas, and if it is maintained in the good tradition established in Zaire then it could provide a sanctuary. However, in his review, Verschuren suggests that, in spite of its great size, the park only 'protects a few' gorillas. There have also been reports of gorillas being killed as a reprisal for raids on crops. A survey is required in the park, along the lines of the one carried out in the Virunga volcanoes.

Elsewhere, efforts are being made to incorporate more of Mt Tsiaberimu into the Virunga National Park. The need for this is said to be acute because clearance of the bamboo forest is increasing, and gorillas are hunted for meat. Schaller estimated the population at between thirty and forty animals in 1959, and in 1975 Verschuren thought that there were still forty gorillas in the area. Both figures are rough estimates, but it is clear that these few gorillas are unlikely to survive unless measures are taken to protect them.

The Kayonza forest in South West Uganda (Fig. 51) contains the only population of eastern lowland gorillas outside Zaire. It lies about twenty-four kilometres north of the Virungas and adjoins Zaire's eastern border. There is no recent information on this area but, in 1959, Schaller estimated that the gorilla population numbered between 120 and 180. The forest was made a reserve (The Impenetrable Central Forest Reserve) in 1932, and when its final boundaries were fixed in 1958 it covered 250 square kilometres. As its name implies, the impenetrable forest is not suitable for use by man, and there is a hope that the gorillas have remained unmolested. A field study in this area would be extremely worthwhile as a basis for future conservation of gorillas in Uganda.

Eastern lowland gorillas are also found to the west of Lake Tanganyika, in the Mtombwe mountains between Mwenga and Fizi. Verschuren has reported that overgrazing has caused extensive erosion in this area. Unless a reserve is created in the Mtombwe mountains it seems unlikely that the gorillas can survive against the encroachments of an enlarging human population.

The western lowland gorilla (*Gorilla g. gorilla*)

This subspecies has provided the vast majority of the captive population and most of the material found in museum collections. Assumed to be the most numerous subspecies, its range is huge, stretching at one time from Nigeria to Zaire (see Fig. 9, page 25). Over twenty years ago Blancou suggested that there might be between 10,000 and 20,000 western lowland gorillas. This figure was calculated from estimates of population density of 0.36 to 0.58 per square kilometre and cannot be considered as more than an approximate guide. Donald Cousins has reviewed much of the information on this subject and I shall refer to his articles as well as to opinions expressed by Critchley, Gartlan, Sabater Pi, Verschuren, Webb and others.

It is likely that the western lowland gorilla is extinct in Zaire and Nigeria, which lie at opposite ends of its former distribution range. Verschuren believes that the animal is effectively extinct in Zaire, although some may occasionally cross the border from Cabinda and Congo (Brazzaville) into what remains of the Mayumbe forest. In Nigeria, gorillas were reported to be in danger of extinction in 1930 and by 1956 were thought to have been virtually wiped out. If there are any left, they are confined to a tiny area on the eastern border adjoining the Takamanda region in West Cameroun. The Takamanda gorillas probably represent an isolated population because the distribution of gorillas in Cameroun follows two gradients of increasing abundance; from north to south and from west to east. An unpublished survey by William Critchley in 1968 indicated that a small pocket of gorillas did survive in Takamanda, but its present status is not known.

In south west Cameroun there are two forest reserves, Dja (526,000 hectares) and Campo (333,000 hectares, Fig. 51). Stephen Gartlan, of the Wisconsin Primate Research Center, carried out a brief survey of Campo in 1974 to assess its potential as a national park. His findings may be applicable to other coastal areas. The reserve is accessible by road and logging companies have built roads inside its boundaries in order to extract timber. Villages exist inside the reserve and forest is cleared by the slash-and-burn technique. Gartlan also found that a good deal of trapping occurred at Campo In the week before he conducted his survey I visited Kribi, which lies just North of Campo. There I met an animal dealer who had five infant and juvenile gorillas awaiting sale to zoos. Gartlan reached the conclusion that Campo would soon become unsuitable for development as a national park. The position of Dja is said to be similar; both areas contain some gorillas but offer little in the way of sanctuary. It is easy to be misled by a line on a map designating an area as a 'Reserve'. Frequently this refers only to an area of controlled exploitation of timber and not an area where animals are protected. In 1974 I worked with Thelma Rowell in the Mbalmayo forest reserve in Cameroun. Talapoin monkeys and prosimians still survived in the area, but gorillas, chimpanzees and most of the larger monkey species had been killed off.

The picture is rather different in south east Cameroun, where vast areas of relatively untouched forest still remain. Timber companies have begun to operate there, and gorillas are occasionally killed if they raid the villagers' crops. In 1976 the Cameroun government designated an area of 400 square kilometres of forest near Mouloundou as a conservation area. These forests, around Lake Lobéké, are very swampy and are probably unsuitable for gorillas.

Bordering Cameroun to the east is the Central African Republic (Fig. 51). The eastward 'increasing gradient of abundance' of gorillas probably terminates somewhere in the region of Sosso and Nola, two towns within sixty kilometres of the Cameroun border. There is no protection for gorillas in the Central African Republic and none of the existing reserves contain them.

In the south, Cameroun borders upon Rio Muni, Gabon and Congo (Brazzaville), all of which are said to contain substantial gorilla populations. In 1964 Sabater Pi estimated that there were 5,000 gorillas in Rio Muni. During 1966–8, however, he did more extensive field work with Clyde Jones and revised the population estimate to 2,000 and finally to a few hundred. Such a dramatic downward revision emphasizes the inaccuracy of preliminary estimates. In the later survey, gorillas were found in four major areas; Mbia Campo, Mobumuom–Monte–Mitra, Mokula and Nkim. One reserve was created in 1953, but it is not thought to contain gorillas. The Mount Alen Partial Reserve is listed as a gorilla sanctuary, but experience suggests that reserves, let alone 'partial' ones, offer little in the way of sanctuary. It seems that up to one third of Rio Muni is exploited by the timber industry and that, though it is illegal, gorillas are often hunted and the young ones are sold to dealers.

Gabon lies to the south and east of Rio Muni. Population pressures are low and gorillas are said to be plentiful, especially in the north east of Gabon. At Makoukou there is a French primate centre and the newly created Ipassa Reserve (15,540 hectares, not marked on Fig. 51). Although excellent field work on prosimians and various monkeys has been carried out at the Makoukou Primate Centre there has been no comparable study of the gorilla. In Gabon gorillas are hunted for food, but it seems that there is no immediate threat of extinction. The authorities discourage trapping, and if gorillas are captured they may be confiscated and handed over to the Makoukou Primate Centre or to the the new International Centre for Medical Research in Franceville.

Several national parks have been established in Gabon. The Okanda National Park (190,000 hectares) and the adjoining Ogue Strict National Reserve (150,000) hectares) lie in a remote and uninhabited area. Gorillas are still to be found in the low coastal forests and are protected in two national parks, the Petit Loango and Wonga-Wongue (82,000 hectares).

The Democratic Republic of the Congo (Congo-Brazzaville) is the last of the countries within the range of the western lowland gorilla. It is a large, thinly populated country in which pressure for land development is not yet intense. Data are very scarce, but gorillas are thought to be plentiful. Gartlan suggests that they are probably most common in the north and east, in approximate continuity with the populations of Cameroun, the Central African Republic and Gabon. In addition there is a national park (Ozala, 110,000 hectares) which was created in 1940 and which is known to contain gorillas. The animals are hunted, mostly with primitive weapons, both for food and in reprisal for plantation destruction. The Mbetis tribe, in particular, have a long tradition of gorilla-hunting, as do the Medjim Mey of Cameroun and the Fang of Rio Muni. Such traditional hunting is not, by itself, a threat to the survival of gorillas.

The overall picture of the western lowland gorilla is one of scanty, unverified information frequently recorded by authors (myself included) who are several removes from its source. Most of the information is out of date and there is a great need for further fieldwork on the animal. In 1974, Gartlan, who is co-ordinator of all the gorilla conservation projects to be sponsored by the International Union for the Conservation of Nature and World Wildlife Fund wrote a strategy for fieldwork on western gorillas. He listed six specific problems for investigation:

1 The western limit of distribution of the subspecies, whether in Nigeria or Cameroun.
2 The delineation of the Takamanda population in Cameroun together with estimates of size, isolation and suitability as a gorilla sanctuary.
3 The northern limits of distribution of the subspecies in Nigeria, Cameroun and the Central African Republic.
4 The eastern limits of distribution in the Central African Republic and Congo-Brazzaville.
5 The southern limits of distribution in Congo-Brazzaville and Gabon.
6 In all these countries; estimates of population size, habitat preference and distribution, hunting and logging pressures and suitability as possible reserves.

Since the scope of the project is huge, Gartlan suggested that a number of field studies should be initiated concurrently and, at that time, four separate projects were planned. As far as I know only one of them, by Miss Julie Webb in Cameroun, has been carried out. One can only hope that the remaining studies will be initiated soon since, without them, there will be insufficient information to develop conservation programmes.

Conservation in captivity

Zoos doubtless play an important role in the gathering of scientific information and in the preservation of endangered species by establishing breeding stocks from which wild sanctuaries might be repopulated. The Indian rhinocerous, European and American bison, Arabian oryx, Przewalski's horse and Père David's deer have all benefited from this type of work. It is not legitimate, however for zoos to plead 'conservation' as their reason for acquiring any species if the creation of a market for that species will cause a serious threat to its survival in the wild. This is now the situation with the gorilla. Attempts to obtain wild specimens usually spring from commercial motives. As Robert Martin pointed out in a recent volume entitled *Breeding Endangered Species in Captivity*: 'Captive breeding should be regarded as an integral part of general conservation programmes and not as a substitute for preservation of populations in the wild state.'

A juvenile gorilla is worth at least 6,000 US dollars. That kind of incentive naturally means that animal dealers will continue to try to obtain gorillas from native hunters so long as zoos are willing to buy them. Under the provisions of the 'Convention on International Trade in Endangered Species', signed by representatives of eighty countries in 1973, the gorilla is listed as a species for which trade is subject to strict regulation and permitted only in exceptional circumstances. Both import and export permits are required and will be issued only when officials are satisfied that the trade will not be detrimental to the survival of the species and that the animals were not poached.

The International Zoo Yearbook of 1979 (Vol. 19) lists a total of 467 gorillas held in 119 collections. If zoos co-operate with each other, then this figure should be adequate for captive breeding programmes and it is unnecessary, as well as undesirable to obtain further stocks from the wild. Of the 467 captive animals, 457 belong to the western lowland subspecies. The ten remaining gorillas, although usually referred to as 'mountain gorillas', are probably all members of the eastern lowland subspecies. It is unlikely that the two eastern subspecies can be successfully conserved in captivity and importing more animals from the wild will not solve the problem. Captive breeding programmes should concentrate on the western subspecies which is held in large numbers and has reproduced successfully in many zoos. There has been a gradual increase in gorilla births such that, in 1978, twenty-two per cent of the animals in zoos had been born in captivity. Twelve collections unfortunately contain adult gorillas of only one sex, but in some cases, as at London Zoo, it has been possible to arrange breeding loans and hence to contribute towards the conservation effort. The encouraging improvement in breeding success among western lowland gorillas cannot be matched among the eastern animals, since there are only ten of them spread between four collections, two of which do not possess both sexes. Only one of these eastern gorillas was born in captivity.

If western lowland gorillas at present in captivity cannot be bred to maintain or increase current stocks, then it is unlikely that further capture of wild specimens would improve the position. Although further capture of eastern lowland or mountain gorillas might enhance a captive breeding effort this would be indefensible in view of the rarity of these animals, particularly the mountain subspecies. Money would be better spent on efforts to study these animals in the wild and in assisting schemes (such as the Fauna Preservation Society's 'Mountain Gorilla Project') to conserve them in a natural habitat. As for the captive animals, if conservation is to be attempted seriously, then it would be best if they were held in a small number of collections. Although the 115 zoos and other institutions which keep western lowland gorillas doubtless find this situation commercially beneficial, it is not likely to be the most effective way to conserve them. Against this argument it must be admitted that to house and maintain a large number of gorillas in one collection would be a very expensive exercise.

Although it is important that gorillas should be bred and studied in captivity, the primary objective must be to ensure their survival in the wild. It should be kept in mind that, apart from conserving the gorilla for scientific reasons, its survival should benefit the countries which it inhabits. Many of those countries are underdeveloped and face serious economic difficulties. If forest parks are administered correctly, they will support a tourist industry, just as the savannah parks of the east and South Africa do. It is also important that the forests themselves should be conserved and ecological studies undertaken. In Rwanda, for instance, the Virunga volcanoes and their forests are the major source of water for the surrounding agricultural community. Ten per cent of the annual rainfall of Rwanda falls on this tiny area and streams flow out of the park into the farmlands. Reduction of the rainforest has inevitably led to a decline in the water supply and if the forest should disappear then a vital water catchment area will be lost. If governments can be persuaded to sympathize with such facts then something may be achieved, even in the face of population pressures and of opposition from timber firms, farmers or mining companies. Fortunately several African countries, Zaire in particular, are acting to conserve their forests and wild animals.

The survival of the gorilla, or its extinction, represents only a tiny fraction of a global problem, namely the decline of the earth's natural resources. In March 1980 a debate on this topic began simultaneously in thirty-two countries as part of a 'World Conservation Strategy' organized by the International Union for the Conservation of Nature. One major issue concerns the destruction of the world's rainforests. The forests of Africa, Asia and South America are being felled at an alarming rate. It has been estimated that, by 1979, forty per cent of all rainforest had been removed and that every year a further 11,000,000 hectares are cut down. This is equivalent to 49.2 acres per minute.

G. Lucas of the Survival Service Commission has pointed out to me that, at that rate of destruction, the entire flora of the Royal Botanical Gardens at Kew would last six minutes. Most of the world's rain forests and many thousands of animals which depend on them may have disappeared by the end of the century. The consequences as far as gorillas and most other primate species are concerned are obvious. The situation is appalling, particularly because, with careful management, the rainforests represent a potentially infinite resource.

I hope that this book has conveyed some idea of the fascination which the gorilla holds for those who study zoology, anthropology, psychology or physiology. I also hope that biologists of the future will still have the opportunity to study this magnificent animal.

Bibliography

A number of references were consulted during the preparation of more than one chapter. In order to save space, these references are listed only once, but are marked with an asterisk.

Chapter 1. Historical Perspective

Akeley, C.E. (1923) *In Brightest Africa*, New York

Bart, Sir W.J. (1833) *The Natural History of Monkeys*, Edinburgh

*Beringe, O. von (1903) *Bericht des Hauptmanns von Beringe über seine Expedition nach Ruanda*

Bontius, J. (1658) *Historiae Naturalis & Medicae Indiae Orientalis* Bk V, Amsterdam

Boulenger, E.G. (1936) *Apes and Monkeys*, Harrap, London

Bowdwich, T.E. (1819) *Mission from Cape Coast to Ashantee*, London

Brehm, A.E. (1895) *Life of Animals* (translated from 3rd German edn), Marquis and Co, Chicago

Buffon, G.L.L. (1766) *Histoire Naturelle, Général et Particulière* Vol. 14, Paris

Burbridge, B. (1928) *Gorilla*, New York

Camper, P. (1779) 'De l'orang-outang, et de quelques autres espèces de singes', in: *Oeuvres de Pierre Camper, qui ont pour object l'histoire naturelle, la physiologie et l'anatomie comparée*, Amsterdam

Chaillu, P. Du (1861) *Exploration and Adventures in Equatorial Africa*, New York

*Chaillu, P. Du (1869) *Stories of the Gorilla Country*, Harper and Bros., New York

Cousins, D. (1972) 'Diseases and injuries of wild and captive gorillas', *Int. Zoo. Yrbook* 12: p 211–18

Cousins, D. (1972) 'Gorillas in captivity, past and present', *Zool. Gart. (N.F.) Leipzig* 42: p 251–81

Denis, A. (1963) *On Safari*, London

Duplaix-Hall, N. (ed.) (1973-5) *Int. Zoo. Yrbook* Vols 13–15, Zool. Soc., London

Garner, R.L. (1896) *Gorillas*, Osgood McIlvaine and Co., London

*Groves, C.P. (1967) 'Ecology and taxonomy of the gorilla' *Nature* 213: p 890–3

Groves, C.P. and Napier, J.R. (1966) 'Skulls and skeletons of *Gorilla* in British collections', *J. Zool. Lond.* 148: p 153–61

Grzimek, B. (1953) 'Die gorillas ausserhalb Africas' *Zool. Gart. (N.F.)* 20: p 173–85

Hediger, H. (1955) *Studies of the Psychology and Behaviour of Captive Animals in Zoos and Circuses*, Butterworth, London

Huxley, T.H. (1904) *Man's Place in Nature*, J.A. Hill and Co., New York

Jarvis, C. (ed.) (1965–8) *Int. Zoo. Yrbook* Vols 5–8, Zool. Soc., London

Jarvis, C. and Morris, D. (eds.) (1961–2) *Int. Zoo. Yrbook* Vols 3–4, Zool. Soc., London

Kawai, M. and Mizuhara, H. (1959) 'An ecological study of the wild mountain gorilla (*G. g. beringei*)', *Primates* 2: p 1–42

Lucas, J. (ed.) *Int. Zoo. Yrbook* Vols 9–11, Zool. Soc., London

Lucas, J. and Duplaix-Hall, N. (eds.) (1972) *Int. Zoo. Yrbook* Vol. 12, Zool. Soc., London

*Merfield, F.G. (1957) *Gorillas were my Neighbours*, London

*Olney, P. (ed.) (1976–9) *Int. Zoo. Yrbook* Vols 16–19, Zool. Soc., London

Owen, R. (1849) 'Osteological contributions to the natural history of the chimpanzees (*Troglodytes*, Geoffroy) including the description of the skull of a large species (*Troglodytes gorilla*, Savage) discovered by Thomas S. Savage, M.D., in the Gaboon country West Africa', *Trans. Zool. Soc. Lond.* 3: p 381–422

Owen, R. (1859) *On the Classification and Distribution of the Mammalia*, Harper and Son, London

Owen, R. (1865) *Memoir on the Gorilla* (Troglodytes gorilla, *Savage*), London

Pigafetta, F. (1591) *A Report of the Kingdom of the Congo and of the Surrounding Countries ; drawn out of the writings and discourses of the Portuguese, Duarte Lopez*, Rome (translated by M. Hutchinson, 1881, London)

Purchas, S. (1613) *Hakluytus posthumus, or Purchas his pilgrims. Contayning a history of the world, in sea voyages and land travells* (4th edn. 1625) Vol 2, London

Raffles, T.S. (1822) 'Descriptive catalogue of a zoological collection, made on account of the honourable East India Company, in the island of Sumatra and its vicinity', *Trans. Linn. Soc. London* 13: p 239–43.

Raven, H.C. (1936) 'In quest of gorillas XII. Hunting gorillas in West Africa' *Sci. Monthly* 47: p 313–34

Reynolds, V. (1967) 'On the identity of the ape described by Tulp, 1641', *Folia Primatol.* 5: p 80–7

Sabater Pi, J. (1966) 'Gorilla attacks against humans in Rio Muni, West Africa', *J. Mammal* 47: p 123–4

Sabater Pi, J. and Groves, C. (1972) 'The importance of higher primates in the diet of the Fang of Rio Muni', *Man* 7: p 239–43

Savage, T.S. (1847) 'Notice of the external characters and habits of a new species of *Troglodytes* (*T. gorilla*, Savage) recently discovered by Dr Savage near the river Gaboon, Africa,' *Proc. Boston Soc. Nat. Hist.* 5: p 245–7

Savage, T.S. and Wyman, J. (1843–4) 'Observations on the external habits and characters of the *Troglodytes niger*, and on its organization', *Boston J. Nat. Hist.* 4: p 362–76 and 377–86

Savage, T.S. and Wyman, J. (1847) 'Notice of the external characters and habits of *Troglodytes gorilla*, a new species of orang from the Gaboon River, osteology of the same', *Boston J. Nat. Hist.* 5: p 417–41

Sweden, Prince William of (1921) *Among Pigmies and Gorillas*, Dutton and Co., New York

Tulp, N. (1641) *Nicolai Tulpii, Amstelredamensis, Observationum Medicarum Libritres*, Amsterdam

Tyson, E. (1699) *Orang-outang sive Homo sylvestris ; or the anatomy of a Pigmie compared with that of a monkey, an Ape and a Man*, London

Vosmaier, A. (1778) 'Description de l'espèce de singe aussi singulier que très rare, nommé orang-outang, de l'île de Borneo', in *Description d'un Receuil Exquis d'animaux rares, Consisten en Quadrupèdes, Oiseaux et Serpents*, Amsterdam.

Yerkes, R.M. (1927) 'The mind of a gorilla', *Genet. Psychol. Monogr.* 2: p 1–193

Yerkes, R.M. (1951) 'Gorilla census and study' *J. Mammal*, 32: p 429–36

*Yerkes, R.M. and Yerkes, A.W. (1929) *The Great Apes: A Study of Anthropoid Life*, Yale Univ. Press, New Haven

Chapter 2. Classification and Distribution

Alix, E. and Bouvier, A. (1877) 'Sur un nouvel anthropoide (*Gorilla mayema*) provenant de la region du Congo', *Bull Soc. Zool. Fr.* p 488–90, Paris

Bearder, S.K. and Martin, R.D. (1979) 'The social organization of a nocturnal primate revealed by radio tracking', in *A Handbook of Biotelemetry and Radiotracking*, Amlaner, C.J. and MacDonald, D.W. (eds.) p 633–48, Pergamon Press, Oxford and New York

Blancou, L. (1951) 'Notes sur les Mammifères de l'equator africain français. Le Gorilla', *Mammalia* 15: p 143–56

Casimir, M.J. (1975) 'Some data on the systematic position of the eastern gorilla population of the Mt. Kahuzi region (Republic du Zaire)', *Z. Morphol Anthropol.* 66: p 188–201

Charles-Dominique, P. (1977) *Ecology and Behaviour of Nocturnal Primates*, Duckworth, London

Clark, W.E. Le Gros (1959) *The Antecedents of Man*, Edinburgh Univ. Press, Edinburgh

Coolidge, H.J. Jr. (1929) 'A revision of the genus *Gorilla*', *Mem Mus. Comp. Zool. Harv.* 50: p 291–381

Corbet, G.B. (1967) 'Nomenclature of the 'eastern lowland gorilla'', *Nature* 215: p 1171–2

Duckworth, W.L.H. (1898) 'Note on an anthropoid ape', *Proc. Zool. Soc. Lond.* p 989–94

Duckworth, W.L.H. (1899) 'Further note on specific differences in the anthropoid apes', *Proc. Zool. Soc. Lond.* 1: p 312–44

*Dunn, F.L. (1966) 'Patterns of parasitism in primates: phylogenetic and ecological interpretations, with particular reference to the hominoidea', *Folia Primatol* 4: p 329–45

Elliott, D.G. (1913) *A Review of the Primates*, Vol. 3, Amer. Mus. Nat. Hist., New York

*Emlen, J.T. and Schaller, G.B. (1960) 'Distribution and status of the mountain gorilla (*Gorilla gorilla beringei*) – 1959,' *Zoologica* 45: p 41–52

Forbes, H.O. (1897) *A Handbook to the Primates*, Vol. 2, Edward Lloyd Ltd., London

Frechkop, S. (1943) 'Mammifères' *Explor. Parc. Natn. Albert Miss. S. Frechkop* No. 1: p 1–186

Goodman, M. (1964) 'Man's place in the phylogeny of the primates as reflected in serum proteins', in *Classification and Human Evolution* (Washburn, S.L., ed.) p 204–34, Methuen and Co., London

Grant, V. (1963) *Origin of Adaptations*, Columbia Univ. Press, New York

Gregory, W.K. and Raven, H.C. (1937) *In Quest of Gorillas*, New Bedford

Groves, C.P. (1970) *Gorillas*, Arco Publ. Co., New York

Groves, C.P. (1970) 'Population systematics of the gorilla', *J. Zool. Lond.* 161: p 287–300

Groves, C.P. (1971) 'Distribution and place of origin of the gorilla', *Man* 6: p 44–51

Groves, C.P. (1972) 'Systematics and phylogeny of gibbons', in *Gibbon and Siamang*, Vol 1 (Rumbaugh, D.M., ed.) p 1–89, Karger, Basel

Haddow, A.J. and Ross, R.W. (1950) 'A critical review of Coolidge's measurements of gorilla skulls', *Proc. Zool. Soc. Lond.* 121: p 43–54

Jantschke, F. (1975) 'The maintenance and breeding of pigmy chimpanzees', in *Breeding Endangered Species in Captivity* (Martin, R.D., ed.) p 245–51, Academic Press, London

Jones, C. and Sabater Pi, J. (1971) 'Comparative ecology of *Gorilla gorilla* (Savage and Wyman) and *Pan troglodytes* (Blumenbach) in Rio Muni', *Bibl. Primatol* No. 13, Karger, Basel

Klinger, H.P., Hamerton, J.L., Mutton, D. and Lang, E.M. (1964) 'The chromosomes of the Hominoidea', in *Classification and Human Evolution* (Washburn, S.L., ed.) p 235–42, Methuen and Co., London

Koenigswald, G.H.R. von (1968) 'The phylogenetic position of the Hylobatinae', in *Taxonomy and Phylogeny of Old World Primates with References to the Origin of Man* (Chiarelli, B. ed.) p 271–6, Rosenberg and Sellier, Torino

*Kortlandt, A. (1972) *New Perspectives on Ape and Human Evolution*, Stitchting voor psychobiologie, Amsterdam

Matschie, P. (1903) 'Uber einen Gorilla aus Deutsch-Ostafrica', *Sitzber Ges. Naturf. Fr. Berl.* p 253–9

Matschie, P. (1905) 'Merkwürdige Gorilla-schädel aus Kamerun' *Sitzber Ges. Naturf. Fr. Berl.* p 279–83

Mivart, St George (1873) 'On *Lepilemur* and *Cheirogaleus* and on the zoological rank of the *Lemuroidea*', *Proc. Zool. Soc. Lond.* p 484–510

Mivart, St George (1873) *Man and Apes, an Exposition of Structural Resemblances and Differences, Bearing upon Questions of Affinity and Origin*, London

Moody, P.A. (1962) *Introduction to Evolution*, Harper and Bros., New York

Moreau, R.E. (1954) 'The distribution of African evergreen forest birds' *Proc. Linn. Soc. Lond.* 165: p 35–46

*Napier, J.R. and Napier, P.H. (1967) *A Handbook of Living Primates*, Academic Press, London

Poirier, F.E. (1974) *In Search of Ourselves*, Burgess Publ. Co. Minneapolis, Minnesota

Rosen, S.I. (1974) *An Introduction to the Primates, Living and Fossil*, Prentice-Hall, New Jersey.

*Sarich, V.M. and Wilson, A.C. (1973) 'Generation time and genomic evolution in primates', *Science* 179: p 1144–6

*Schaller, G.B. (1963) *The Mountain Gorilla: Ecology and Behavior*, Univ. Chicago Press, Chicago

Schouteden, H. (1927) 'Gorille de Walikale', *Rev. Zool. Afr.* 15: p 47

Schultz, A.H. (1934) 'Some distinguishing characters of the mountain gorilla' *J. Mammal* 15: p 51–61

*Schultz, A.H. (1968) 'The recent hominoid primates' in *Perspectives on Human Evolution* (Washburn, S.L. and Jay, P.C., eds.) p 122–95, Holt, Rinehart and Winston, New York

183

*Schultz, A.H. (1969) *The Life of Primates*, Weidenfeld and Nicolson, London
Simpson, G.G. (1964) 'The meaning of taxonomic statements', in *Classification and Human Evolution* (Washburn, S.L. ed.) p 1–31, Methuen and Co., London
Vogel, C. (1961) 'Zur systematische Untergliederung de Gattung *Gorilla* anhand von Untersuchungen der Mandibel', *Z. Saugetierk* 26: p 65–76

Chapter 3. Structure and Function

Aston, E.H. (1954) 'Age changes in the bodily proportions of some apes', *Proc. Zool. Soc. Lond.* 174: p 587–94
Atkinson and Elftman, H. (1950) 'Female reproductive system of the gorilla', in *The Anatomy of the Gorilla*: The Henry Cushier Raven Memorial Volume, p 205–7, Columbia Univ. Press, New York
Badoux, D.M. (1974) 'An introduction to biomechanical principles in primate locomotion and structure', in *Primate Locomotion* Farish O. Jenkins, Jr. (ed.) p 1–43, Academic Press, New York
Beddard, F.E. (1909) *Mammalia*, Macmillan and Co., London
Cousins, D. (1972) 'Body measurements and weights of wild and captive gorillas' *Zool. Gart. (N.F.)* Leipzig 41: p 261–77
Duckworth, W.L.H. (1915) *Morphology and Anthropology Vol. 1* (2nd edn), Cambridge Univ. Press, Cambridge
Elftman and Atkinson, W.B. (1950) 'The abdominal viscera of the gorilla,' in *The Anatomy of the Gorilla; The Henry Cushier Raven Memorial Volume*, p 197–201, Columbia Univ. Press, New York
Ellis, R.A. and Montagna, W. (1962) 'The skin of the primates VI. The skin of the gorilla (*Gorilla gorilla*)', *Amer J. Phys. Anthrop.* 20: p 79–94
Erikson, G.E. (1963) 'Brachiation in New World monkeys and in anthropoid apes', *Symp. Zool. Soc. Lond.* 10: p 135–63
Flower, W.H. (1872) 'Lectures on the comparative anatomy of the organs of digestion of the Mammalia: Lecture III', *Med. Times and Gazette* 1: p 334–7 and 392–4
Gijzen, A. and Tijskens (1971) 'Growth in weight of the lowland gorilla, *Gorilla g. gorilla*, and of the mountain gorilla, *Gorilla g. beringei*' *Int. Zoo. Yrbk.* II: p 183–93
*Hall-Craggs, E.C.B. (1962) 'The testis of *Gorilla gorilla beringei*', *Proc. Zool. Soc. Lond.* 139: p 511–14
Hofer, H. (1972) 'Über den jesang des Orang-Utan (*Pongo pygmaeus*)', *Zool. Gart. (N.F.) Leipzig* 41: p 299–302
Hooton, E. (1942) *Man's Poor Relations*, Doubleday, Doran and Co. Inc., New York.
Hosokama, H. and Kamiya, T. (1961–2) 'Anatomical sketches of the visceral organs of the mountain gorilla (*Gorilla gorilla beringei*)', *Primates* 3: p 1–28
Keleman, G. (1961) 'Anatomy of the larynx as a vocal organ: evolutionary aspects', *Logos 4* : p 46–55
Marzke, M.W. (1971) 'Origin of the human hand' *Amer. J. Phys. Anthrop.* 34: p 61–84
Osman-Hill, W.C. and Matthews, H.L. (1949) 'The male external genitalia of the gorilla, with remarks on the os penis of other Hominoidea', *Proc. Zool. Soc. Lond.* 119: p 363–79

Raven, H.C. *The Anatomy of the Gorilla*, The Henry Cushier Raven Memorial Volume, Columbia Univ. Press, New York

Roberts D. 'Structure and function of the primate scapula', in *Primate Locomotion* (Farish O. Jenkins Jr. ed.) p 171–200, Academic Press, New York

Rose, M.D. (1974) 'Ischial tuberosities and ischial callosities' *Amer. J. Phys. Anthrop.* 40: p 375–83

Schultz, A.H. (1936) 'Characters common to higher primates and characters specific for man' *Quart. Rev. Biol.* 11: p 259–83.

Schultz, A.H. (1950) 'Morphological observations on gorillas' in *The Anatomy of the Gorilla*, The Henry Cushier Raven Memorial Volume, p 227–53 Columbia Univ. Press, New York

Schultz, A.H. (1961) 'Vertebral column and thorax', *Primatologia* 4 Liefer 5: p 1–66.

Schultz, A.H. (1963) 'Relations between the lengths of the main parts of the foot skeleton in primates', *Folia Primatol* 1: p 150–71

Sonntag, C.F. (1924) *The Morphology and Evolution of the Apes and Man*, John Bale, Sons & Danielsson, London

Steiner, P. (1954) 'Anatomical observations on a *Gorilla gorilla*', *Amer. J. Phys. Anthrop.* 12: p 145–79

Straus, W.L. Jr. (1950) 'The microscopic anatomy of the skin of the gorilla', in *The Anatomy of the Gorilla: The Henry Cushier Raven Memorial Volume* p 213–21, Columbia Univ Press, New York

*Thomas, W.D. (1958) 'Observations on the breeding in captivity of a pair of lowland gorillas' *Zoologica* 43: p 95–104

Tuttle, R.H. (1967) 'Knuckle-walking and the evolution of hominoid hands' *Amer, J. Phys. Anthrop* 26: p 171–206

Tuttle, R.H. and Basmajian, J.V. (1974) 'Electromyography of forearm musculature in *Gorilla* and problems related to knuckle-walking', in *Primate Locomotion* (Farish O. Jenkins ed.) p 293–347, Academic Press: New York

Tuttle, R. and Beck B.B. (1972) 'Knuckle-walking hand postures in an orang-utan (*Pongo pygmaeus*)', *Nature* 236: p 33–4

Washburn, S.L. (1957) 'Isohial callosities as sleeping adaptations', *Amer. J. Phys. Anthrop.* 15: p 269–76

Wislocki, G.B. (1932) 'On the female reproductive tract of the gorilla, with a comparison of that of other primates', *Contrib. Embyol. Carnegie Inst.* 23: p 165

Wislocki, G.B. (1942) 'Size, weight and histology of the testes in the gorilla', *J. Mammal* 23: p 281–7

Chapter 4. How Close to Man?

Benveniste, R.E. and Todaro, G.J. (1976) 'Evolution of type C viral genes: evidence for an Asian origin of man', *Nature* 261: p 101–8

Chiarelli, B. (1968) 'Caryological and hybridological data for the taxonomy and phylogeny of the Old World primates', in: *Taxonomy and Phylogeny of the Old World Primates with References to the Origin of Man* (Chiarelli B, ed.) p 151–86, Rosenberg and Sellier, Torino

Chiarelli, B. (1973) *Evolution of the Primates*, Academic Press, London

Day, M.H. and Wood, B.A. (1969) 'Hominoid tali from East Africa', *Nature*, 222: p 591

Garnham, P.C.C. (1973) 'Distribution of malaria parasites in primates, insectivores and bats', *Symp. Zool. Soc. Lond.* 33: p 377–404

Geiman and Meagher (1967) 'Susceptibility of a New World monkey to *Plasmodium falciparum* from man', *Nature* 215: p 437

Goodman, M. (1968) 'Phylogeny and taxonomy of the catarrhine primates from immunodiffusion data I. A review of the major findings', in *Taxonomy and Phylogeny of Old World Primates with References to the Origin of Man* (Chiarelli, B., ed) p 95–107, Rosenberg and Sellier, Torino

Goodman, M. (1973) 'The chronicle of primate phylogeny contained in proteins', *Symp. Zool. Soc. Lond.* 33: p 339–75

Goodman, M. and Moore, G.W. (1974) 'Phylogeny of haemoglobin' *Syst. Zool.* 32: p 508–32

Gregory, W.H. (1934) *Man's Place Among the Anthropoids*, Clarendon Press, Oxford

Gregory, W.H. and Hellman, M. (1926) 'The dentition of *Dryopithecus* and the origin of man', *Anthrop. papers. Amer. Mus. Nat. Hist.* Vol. 28, part I

Hartman, C.G. (1939) 'The use of the monkey and ape in studies of human biology with special reference to primate affinities', *Amer. Naturalist* 73: p 139–55

Hurzeler, J. (1954) 'Contributions a l'odontologie et a la phylogenie du genre *Pliopithecus* Gervais', *Ann Paleont.* 40: p 5

Ihering, H. von (1891) 'On the ancient relations between New Zealand and South America', *Trans. Proc. New Zealand Inst.* 24: p 431–45

King, M.C. and Wilson, A.C. (1975) 'Evolution at two levels in humans and chimpanzees', *Science* 188: p 107–16

Kortlandt, A. and Zon, J.C.J. van (1969) 'The present state of research on the dehumanization hypothesis of African ape evolution', in *Proc. 2nd Int. Congr. Primatol.* Vol. 3, p 10–13, Karger, Basel

Kuhn, H.J. (1968) 'Parasites and the phylogeny of catarrhine primates' in *Taxonomy and Phylogeny of Old World Primates with References to the Origin of Man* (Chiarelli, B., ed.) p 187, Rosenberg and Sellier, Torino

Le Gros Clark, W. and Leakey, L.S.B. (1951) 'The Miocene Hominoidea of East Africa', *Fossil Mammals of Africa* no. 1 Brit. Mus. Nat. Hist., London

Le Gros Clark, W. and Thomas, D.P. (1951) 'Associated jaws and limb bones of *Limnopithecus maccinesi*', *Fossil Mammals of Africa* no. 3, Brit. Mus. Nat. Hist., London

Lewis, G.E. (1934) 'Preliminary notice of new, man-like apes from India', *Amer. J. Sci.* 27: p 161

Mayr, E. (1963) *Animal Species and Evolution*, Harvard Univ. Press, Cambridge, Mass.

Millar, D.A. (1978) 'Evolution of primate chromosomes: man's closest relative may be the gorilla, not the chimpanzee' *Science* 198: p 1116–24

Nuttall, G.H.F. (1904) *Blood Immunity and Blood Relationship*, Cambridge University Press, Cambridge

Pilbeam, D.R. (1969) 'Tertiarry pongidae of East Africa: Evolutionary relationships and taxonomy, *Peabody Mus. Bull.* (*Yale*) no. 31

Pilbeam, D.R. (1972) *The Ascent of Man*, Macmillan, New York

Pilgrim, G.E. (1915) 'New Siwalik primates and their bearing on the question of the evolution of man and the Anthropoidea', *Rec. Geol. Survey India* XLV, part I: p 1–74

Sarich, V.M. (1968) 'The origin of the hominids: an immunological approach', in

Perspectives on Human Evolution Vol. I (Washburn, S.L. and Jay, P.C., eds.) p 94, Holt, Rinehart and Winston, New York

Sarich, V.M. and Wilson, A.C. (1968) 'Immunological time scale for human evolution', *Science* 158: p 1200

Schon, M.A. and Ziemer, L.K. (1973) 'Wrist mechanisms and locomotive behaviour of *Dryopithecus (Proconsul) africanus*', *Folia Primatol.* 20: p 1–11

Simons, E.L. and Pilbeam, D.R. (1965) 'Preliminary revision of the Dryopithecinae (Pongidae, Anthropoidea)' *Folia Primatol*, 3: p 81

Smith, R.J. and Pilbeam, D.R. (1980) 'Evolution of the orang-utan' *Nature* 284: p 447–8

Tijio, J.H. and Levan, A. (1956) 'The chromosome number of man', *Hereditas* 42: p 1–6

Zihlman, A.L., Cronin, J.E., Cramer, D.L. and Sarich, V.M. (1978) 'Pigmy chimpanzee as a possible prototype for the common ancestor of humans, chimpanzees and gorillas', *Nature* 275: p 744–6

Chapter 5. Senses and Intelligence

Albrecht, H. and Dunnett, S.C. (1971) *Chimpanzees in Western Africa*, Piper, München

Alcock, J. (1972) 'The evolution of the use of tools by feeding animals', *Evolution.* 26: p 464–73

Benchley, B.J. (1942) *My Friends the Apes*, Little, Brown and Co, Boston, Mass.

Finch, G. (1941) 'The solution of patterned string problems by chimpanzees', *J. Comp. Psychol.* 32: p 83–90

Fischer, G.J. (1962) 'The formation of learning sets in young gorillas', *J. Comp. Physiol. Psychol.* 55: p 924

Gardner, B.T. and Gardner, R.A. (1971) 'Two-way communication with an infant chimpanzee', in *Behaviour of Non-Human Primates* Vol. 4 (Schrier, A.M. and Stollnitz, F. eds.) p 117–84, Academic Press, New York and London

Gardner, R.A. and Gardner, B.T. (1969) 'Teaching sign language to a chimpanzee', *Science* 165: p 664–72

Gervais, P. (1854) *Histoire Naturelle des Mammifères* Curmer, Paris

Gildakas-Brindamour, B. (1975) 'Orang-utans, Indonesia's "people of the forest"', *Nat. Geogr.* 148: p 444–73

Groves, C.P. and Humphrey, N.K. (1973) 'Asymmetry in gorilla skulls: Evidence of lateralized brain function', *Nature* 244: p 53–4

Hall, K.R.L. and Schaller, G.B. (1964) 'Tool-using behavior of the Californian sea otter' *J. Mammal.* 45: p 287–98

Harlow, H.F. (1949) 'The formation of learning sets' *Psychol. Rev.* 56: p 51–65

Hornaday, W.T. (1922) *The Minds and Manners of Wild Animals*, Scribner, New York

Jones, C. and Sabater Pi, J. (1968) 'Sticks used by chimpanzees in Rio Muni, West Africa', *Nature* 223: p 100–1

Kortlandt, A. (1962) 'Chimpanzees in the wild', *Sci Amer.* 206: p 128–38

Kortlandt, A. (1967) 'Experimentation with chimpanzees in the wild, in *Progress in Primatology* (Stark, D., Scheider, R. and Kuhn, H.J. eds.) p 208–224, Fischer, Stuttgart

Kellogg, W. N. and Kellogg, L. A. (1933) *The Ape and the Child*, McGraw Hill, New York

Köhler, W. (1925) *The Mentality of Apes*, Harcourt and Brace, New York

Kruuk, H. (1972) *The Spotted Hyena*, Univ. Chicago Press, Chicago

Lack, D. (1947) *Darwin's Finches*, Cambridge Univ. Press, Cambridge

Lawick-Goodall, J. van (1973) 'The behavior of chimpanzees in their natural habitat', *Amer. J. Psychiat.* 130: p 1–12

Lawick-Goodall, J. van (1973) 'Cultural elements in a chimpanzee community', *Symp. IVth Int. Cong. Primat. Vol. I Precultural Primate Behavior* (Menzel, E. W. ed.) p 144–84, Karger, Basel

Lawick-Goodall, J. van and Lawick-Goodall, H. van (1966) 'Use of tools by Egyptian vultures', *Nature* 212: p 1468–9

McGrew, W. C. and Tutin, E. G. (1973) 'Chimpanzees' tool use in dental grooming', *Nature* 241: p 477–8

*MacKinnon, J. (1974) 'The ecology and behaviour of wild orang-utans (*Pongo pygmaeus*)', *Anim. Behav.* 22: p 3–74

Menzel, E. W. (1972) 'Spontaneous invention of ladders in a group of young chimpanzees', *Folia Primatol.* 17: p 87–106

Menzel, E. W. (1973) 'Further observation on the use of ladders in a group of young chimpanzees', *Folia Primatol.* 19: p 450–7

Menzel, E. W., Davenport, R. K. and Rogers, C. M. (1970) 'The development of tool-using in wild born and restriction-reared chimpanzees' *Folia Primatol.* 12: p 273–83

Morris, D. (1961) *The Biology of Art*, Knopf, New York

Premack, D. (1971) 'Language in chimpanzees' *Science* 172: p 808–22

Premack, D. (1971) 'On the assessment of language competence in the chimpanzee', in *Behavior of Non-Human Primates* Vol. 4 (Schrier, A. M. and Stollnitz, F., eds.), Academic Press, New York

Reisen, A. H. Greenberg, B., Granston, A. S. and Fanty, R. L. (1953) 'Solutions of patterned string problems by young gorillas', *J. Comp. Physiol. Psychol.* 46: p 19–22

Rensch, B. (1973) 'Play and art in apes and monkeys', in *Symp. IVth Int. Congr. Primat. Vol I. Precultural Primate Behavior* (Menzel, E. W., ed.) p 102–23, Karger, Basel

Reynolds, V. and Reynolds, F. (1965) 'Chimpanzees of the Budongo Forest', in *Primate Behaviour* (DeVore, I., ed.) p 368–424, Holt, Rinehart and Winston, New York

*Rodman, P. S. (1973) 'Population composition and adaptive organization among captive orang-utans of the Kutai Reserve', in *Comparative Ecology and Behaviour of Primates* (Crook, J. H. and Michael, R. P. eds.) p 171–209, Academic Press, London

Rumbaugh, D. M. (1970) 'Learning skills of anthropoids' in *Primate Behavior* Vol I. (Rosenblum, L. A. ed.) p 2–70, Academic Press, New York

Rumbaugh, D. M. (1974) 'Comparative primate learning and its contribution to understanding development, play, intelligence and language', in *Perspectives in Primate Biology* Vol. 9 (Chiarelli, B., ed.) p 253–81, Plenum, New York

Rumbaugh, D. M. and Gill, T. (1973) 'The learning skills of the great apes', *J. Human Evol.* 2: p 171–9

Rumbaugh, D. M., Gill, T. V. and Wright, S. C. (1971) 'Readiness to attend to visual foreground cues: a species – associated perceptual characteristic', *J. Human Evol.* 2: p181–8

*Sabater Pi, J. (1977) 'Contribution to the study of alimentation of lowland gorillas in the natural state, in Rio Muni, Republic of Equatorial Guinea (West Africa)', *Primates* 18: p183–204

Schaller, G.B. (1972) *The Serengeti Lion*, Univ. Chicago Press, Chicago.

Schiller, P.H. (1951) 'Figural preferences in the drawing of a chimpanzee', *J. Comp. Physiol. Psychol.* 44: p101–11

Struhsaker, T.T. and Hunkeler, P. (1971) 'Evidence of tool-using by chimpanzees in the Ivory Coast', *Folia Primatol.* 15: p212–19.

Sugiyama, Y. (1969) 'Social behaviour of chimpanzees in the Bundongo forest, Uganda', *Primates* 10: p197–225

Suzuki, A. (1966) 'On the insect-eating habits among wild chimpanzees living in the savannah woodland of western Tanzania', *Primates* 7: p481–7

Teleki, G. (1973) 'The omnivorous chimpanzee', *Sci. Amer.* 228: p32–42

Tigges, J. (1963) 'On colour vision in gibbon and orang-utan', *Folia Primatol.* I: p188–98.

Yerkes, R.M. (1916) 'The mental life of monkeys and apes; a study in ideational behavior', *Behav. Monogr.* 3: p1–145

Yerkes, R.M. (1927) 'The mind of a gorilla', *Genetic Psychol. Monogr.* 2: p1–193

Yerkes, R.M. (1927) 'The mind of a gorilla Part II Mental development', *Genetic Psychol. Monogr.* 2: p375–551

Yerkes, R.M. (1928) 'The mind of a gorilla Part III Memory', *Comp. Psychol. Monogr.* 5: p1–87

Chapter 6. Behaviour and Ecology

Altmann, S. (ed.) (1965) *Japanese Monkeys*, Edmonton

Baldwin, L.A. and Teleki, G. (1973) 'Field research on chimpanzees and gorillas: an historical, geographical and bibliographical listing', *Primates* 14: p315–30

Baldwin, L.A. and Teleki, G. (1974) 'Field research on gibbons, siamangs, and orang-utans: an historical, geographical and bibliographical listing', *Primates* 15: p365–76

Bernstein, I.S. (1962) 'Response to nesting materials of wild-born and captive-born chimpanzees', *Anim. Behav.* 10: p1–6

Bernstein, I.S. (1970) 'Primate status hierarchies', in *Primate Behavior* Vol. I (Rosenblum, L.A., ed.) p71–109, Academic Press, London and New York

Bingham, H.C. (1932) 'Gorillas in a native habitat', *Carnegie Inst. Wash. Publ.* 426: p1–66

Blower, J. (1956) 'The mountain gorilla', *Uganda Wildlife and Sport* 1: p41–52

Bolwig, N. (1959) 'A study of the nests built by mountain gorilla and chimpanzee' *S. Afr. J. Sci.* 55 (II): p286–91

Burt, W.H. (1943) 'Territoriality and home range concepts as applied to mammals', *J. Mammal* 24: p346–52

Carpenter, C.R. (1940) 'A field study in Siam of the behavior and social relations of the gibbon', *Comp. Psychol. Monogr.* 16: p1–212

Carpenter, C.R. (1965) 'The howlers of Barro Colorado Island', in *Primate Behavior* (DeVore, I., ed.) p250–91, Holt, Rinehart and Winston, New York

Casimir, M.J. (1975) 'Feeding ecology and nutrition of an eastern gorilla group in the Mt Kahuzi region (Republique du Zaire)', *Folia Primatol.* 24: p 81–136

Casimir, M.J. (1979) 'An analysis of gorilla nesting sites of the Mt Kahuzi region (Zaire)', *Folia Primatol.* 32: p 290–308

Casimir, M.J. and Butenandt, E. (1973) 'Migration and core area shifting in relation to some ecological factors in a mountain gorilla group (*Gorilla gorilla beringei*) in the Mt Kahuzi region (Republique du Zaire)', *Z. Tierpsychol.* 33: p 514–22

Chalmers, N. (ed.) (1974) 'Current primate field studies', *Primate Eye* no. 3 (suppl.)

Chivers, D. (1974) 'The siamang in Malaya', *Contribs. to Primatol* no. 4, Karger, Basel.

Chorley, C.W. (1928) 'Notes on Uganda gorillas seen during a visit to Mt Sabinio, Christmas, 1927', *Proc. Zool. Soc. Lond.* 98: p 267–8

Derscheid, J.M. (1927) 'Notes sur les gorilles des volcans de Kivu (Parc National Albert)', *Ann. Soc. Royal Zool. Belg.* 58: p 149–59

Dixson, A.F. (1977) 'Observations on the displays, menstrual cycles and sexual behaviour of the "Black Ape" of Celebes (*Macaca nigra*)', *J. Zool. Lond.* 182: p 63–84

Dixson, A.F., Scruton, D.M. and Herbert, J. (1975) 'Behaviour of the talapoin monkey (*Miopithecus talapoin*) studied in groups, in the laboratory', *J. Zool. Lond.* 176: p 177–210

Donisthorpe, J. (1958) 'A pilot study of the mountain gorilla (*Gorilla gorilla beringei*) in South West Uganda, February to September 1957', *S. Afr. J. Sci.* 54: p 195–217

Ellefson, J.O. (1968) 'Territorial behavior in the common, white-handed gibbon, *Hylobates lar* Linn', in *Primates, Studies in Adaptation and Variability* (Jay, P.C., ed.) p 180–199, Holt, Rinehart and Winston, New York

Elliott, R.C. (1976) 'Observations on a small group of mountain gorillas (*Gorilla gorilla beringei*)', *Folia Primatol.* 25: p 12–24

Emlen, J.T. (1960) 'Current field studies of the mountain gorilla', *S. Afr. J. Sci.* 56(4): p 88–9

Emlen, J.T. (1962) 'The display of the gorilla', *Proc. Amer. Phil. Soc.* 106: p 516–19

Emlen, J.T. and Schaller, G.B. (1960) 'In the home of the mountain gorilla', *Anim. Kingdom* 63: p 98–108

Fiennes, R. (1967) *Zoonoses of Primates*, Weidenfeld and Nicolson, London

Fossey, D. (1970) 'Making friends with mountain gorillas', *Nat. Geogr.* 137(1): p 48–67

Fossey, D. (1971) 'More years with mountain gorillas', *Nat. Geogr.* 140: p 574–85

Fossey, D. (1972) 'Vocalizations of the mountain gorilla', *Anim. Behav.* 20: p 36–53

Fossey, D. (1974) 'Observations on the home range of one group of mountain gorillas (*Gorilla gorilla beringei*)', *Anim. Behav.* 22: p 568–81

Fossey, D. and Harcourt, A.H. (1977) 'Feeding ecology of free-ranging mountain gorilla (*Gorilla gorilla beringei*)', in *Primate Ecology: Studies of Feeding and Ranging Behaviour in Lemurs, Monkeys and Apes* (Clutton-Brock, T.H., ed.) p 415–47, Academic Press, London

Gartlan, J.S. (1964) 'Dominance in East African monkeys' *Proc. E. Afr. Acad.* 2: p 75–9

Goodall, J. (1965) 'Chimpanzees of the Gombe Stream Reserve' in *Primate Behavior* (DeVore, I., ed.) p 425–73, Holt, Rinehart and Winston, New York

Goodall, A.G. (1977) 'Feeding and ranging behaviour of a mountain gorilla (*Gorilla gorilla beringei*) in the Tshibindi-Kahuzi region (Zaire)', in *Primate Ecology:*

Studies of Feeding and Ranging Behaviour in Lemurs, Monkeys and Apes (Clutton-Brock, T.H. ed.) p 449–79, Academic Press, London

*Groom, A.F.G. (1973) 'Squeezing out the mountain gorilla' *Oryx* 12: p 207–15

Hall, K.R.L. and DeVore, I. (1965) 'Baboon social behavior', in *Primate Behavior* (DeVore, I. ed.) p 53–110, Holt, Rinehart and Winston, New York

*Harcourt, A.H. and Groom, A.F.G. (1972) 'Gorilla census', *Oryx* 5: p 355–63

Harcourt, A.H., Stewart, K.S. and Fossey, D. (1976) 'Male emigration and female transfer in wild mountain gorillas', *Nature* 263: p 226–7

Harrison, B. (1963) *Orang-utan*, Doubleday and Co, New York

Hartmann, R. (1886) *Anthropoid Apes*, Appleton and Co, New York

*Hess, J.P. (1973) 'Some observations on the sexual behaviour of captive lowland gorillas (*Gorilla g. gorilla*)', in *Comparative Ecology and Behaviour of Primates* (Michael R.P. and Crook, J.H., eds.), p 507–81, Academic Press, London

Hladik, C.M. (1977) Chimpanzees of Gabon and chimpanzees of Gombe: some comparative data on the diet', in *Primate Ecology: Studies of Feeding and Ranging Behaviour in Lemurs, Monkeys and Apes* (Clutton-Brock, T.H. ed.) p 481–501, Academic Press, London

Hooff, J.A.R.A.M. van (1967) 'The facial displays of the catarrhine primates' in *Primate Ecology* (Morris, D., ed.) p 7–68, Weidenfeld and Nicolson, London

Huber, E. (1931) *'Evolution of the Facial Musculature and Facial Expression*, John Hopkin's Univ. Press, Baltimore

Itani, J. (1958) 'On the acquisition and propagation of a new food habit in the natural group of the Japanese monkey at Takasaki-Yama' *Primates* 1: p 84–98

Jones, C. and Sabater Pi, J. (1971) 'Comparative ecology of *Gorilla gorilla* (Savage and Wyman) and *Pan Troglodytes* (Blumenbach) in Rio Muni', *Bibl. Primatol* no. 13, Karger, Basel

Kawai, M. and Mizuhara, H. (1959) 'An ecological study of the wild mountain gorilla (*G. g. beringei*)', *Primates* 2: p 1–42

Kortlandt, A. (1962) 'Chimpanzees in the wild', *Sci Amer.* 206: p 128–38

Kummer, H. (1971) *Primate Societies: Group Techniques of Ecological Adaptation*, Aldine-Atherton, Chicago

Lang, E. (1963) *Goma the Gorilla Baby*, Doubleday and Co, New York

Lang, E. (1964) 'Jambo – first gorilla raised by its mother in captivity', *Nat. Geogr.* 125: p 446–53

*Lawick-Goodall, J. van (1968) 'The behaviour of free-living chimpanzees in the Gombe Stream Reserve', *Anim. Behav. Monogr.* 1: p 161–311

Lawick-Goodall, J. van (1968) 'A preliminary report on expressive movements and communication in the Gombe Stream chimpanzees' in *Primates, Studies in Adaptation and Variability* (Jay, P.C., ed.) p 313–74, Holt, Rinehart and Winston, New York

Lawick-Goodall, J. van (1971) *In the Shadow of Man*, Collins, London

Le Boeuf, B.J. and Petrinovich, L.F. (1974) 'Dialects of northern elephant seals (*Mirounga angustirostris*): origins and reliability' *Anim. Behav.* 22: p 656–63

Lequime, M. (1959) 'Sur la piste du gorille', *La Vie des Bêtes* 14: p 7–8

Malbrant, R. and Maclatchy, A. (1949) 'Faune de l'equateur Africain Francais II Mammifères', *Encycl. Biol.* 36: p 1–323

Marler, P. (1965) 'Communication in monkeys and apes' in *Primate Behavior* (DeVore, I., ed.) p 544–84, Holt, Rinehart and Winston, New York

Marler, P. (1976) 'Social organization, communication and graded signals: the chimpanzee and gorilla.', in *Growing Points in Ethology* (Bateson, P.P.G. and Hinde, R.A., eds) p 239–80, Cambridge Univ. Press, Cambridge

Marler, P. and Hobbet, L. (1975) 'Individuality in a long-range vocalization of wild chimpanzees', *Zeitschrift für Tierpsychol.* 38: p 97–109

Marler, P. and Tamura, M. (1962) 'Song dialects in three populations of white-crowned sparrows', *Condor* 63: p 368–77

Marsh, C. W. (1979) 'Female transference and mate choice among Tana River red colobus', *Nature* 281: p 568–69

Mason, W.A. (1968) 'Use of space by *Callicebus* groups', in *Primates Studies in Adaptation and Variability* (Jay, P.C., ed.) p 200–16, Holt, Rinehart and Winston, New York

Meadow, S.R. and Smithnells, R.W. (1974) *Lecture Notes on Paediatrics*, Blackwell Scientific Publ., Oxford

Nissen, H.W. (1931) 'A field study of the chimpanzee', *Comp. Psychol. Monogr.* 8 (1): p 1–122

Osborn, R.M. (1963) 'Observations on the behaviour of the mountain gorilla' *Symp. Zool. Soc. Lond.* 10: p 29–37

Phoenix, C.H. (1973) 'The role of testosterone in the sexual behaviour of laboratory male rhesus', in *Symp. IVth Int. Cong. Primatol.* Vol. 2 *Primate Reproductive Behavior* (Phoenix, C.H., ed.) p 99–122, Karger, Basel

Pitcairn, T.K. (1974) 'Aggression in natural groups of pongids', in *Primate Aggression, Territoriality and Xenophobia* (Holloway, R.L. ed.) p 241–268, Academic Press, New York

Pitman, C.R.S. (1935) 'The gorillas of the Kayonza region, Western Kigesi, S.W. Uganda', *Proc. Zool. Soc. Lond.* 105: p 477–99

Reynolds, V. (1965) *Budongo, an African Forest and its Chimpanzees*, Nat. Hist. Press, New York.

Reynolds, V. (1965) 'Some behavioral comparisons between the chimpanzee and the mountain gorilla in the wild', *Am. Anthropol.* 67: p 691–706

Reynolds, V. and Reynolds, F. (1965) 'Chimpanzees of the Budongo Forest', in *Primate Behavior* (DeVore, I., ed.) p 368–424, Holt, Rinehart and Winston, New York

Richards, P.W. (1952) *The Tropical Rain Forest: An Ecological Study*, Univ. Chicago Press, Chicago

Rijksen, H.D. (1978) *A Field Study on Sumatran Orang Utans (Pongo pygamaeus abelii Lesson 1827) Ecology, Behaviour and Conservation*, Veenman & Zonen, Wageningen

Rowell, T.E. (1962) 'Agonistic noises of the rhesus monkey (*Macaca mulatta*)', *Symp. Zool. Soc. London.* 8: p 91–6

Rowell, T.E. (1967) 'A quantitative comparison of the behaviour of a wild and caged baboon group', *Anim. Behav.* 15: p 499–509

Rowell, T.E. (1972) 'Towards a natural history of the talapoin monkey in Cameroun', *Ann. Fac. Sci.* (Cameroun) 10: p 121–34

Sabater Pi, J. (1960) 'Beitrag zur biologie des flachland Gorillas' *Z.f. Saugetierk* 25 (3): p 133–41

Sabater, Pi, J. and Lassaletta, L. De (1958) 'Beitrag zur Kenntis des flachland Gorillas (*Gorilla gorilla*, Savage and Wyman)' *Z.f. Saugetierk* 23: p 108–14

Schaller, G.B. (1961) 'The orang-utan in Sarawak', *Zoologica* 46: p 73–82

Schaller, G.B. (1964) *The Year of the Gorilla*, Univ. Chicago Press, Chicago

Schaller, G.B. (1965) 'The behavior of the mountain gorilla' in *Primate Behavior* (DeVore, I., ed.) p 324–67, Holt, Rinehart and Winston, New York

Schjelderup-Ebbe, T. (1935) 'Social behavior of birds', in *A Handbook of Social Psychology* (Murchison, C.A., ed.) p 947–72, Clark Univ. Press, Worcester, Mass.

Schultz, A.H. (1956) 'The occurrence and frequency of pathological and teratological conditions and of twinning among non-human primates', in *Primatologica* Vol. 1, p 965–1014

Schultz, M. and Stark, D. (1977) 'Neue Beobachtungen und Uberlegungen zur Pathologie des Primatenschadels', *Folia Primatol.* 28: p 81–108

Stecker, R.M. (1958) 'Osteoarthritis in the gorilla Description of a skeleton with involvement of knee and spine', *Lab. Invest.* 7: p 445–57

Stefanick, M. (Unpublished manuscript) 'An observational study of six captive, lowland gorillas (*Gorilla gorilla gorilla*)'

Stefanick, M. (Unpublished manuscript) 'An analysis of a few communicatory signals in six captive, lowland gorillas (*Gorilla gorilla gorilla*)'

Teleki, G. (1973) *The Predatory Behavior of Wild Chimpanzees*, Bucknell Univ. Press, Lewisburg

Tinbergen, N. (1951) *The Study of Instinct*, Clarendon Press, Oxford

Tokuda, K. (1961–1962) 'A study on sexual behavior in the Japanese monkey troop', *Primates* 3: p 1–10

Webb, J.C. (Unpublished manuscript) 'Summary of additional data gathered in Cameroun on western lowland gorillas *Gorilla g. gorilla* June–August, 1974'

Wrangham, R.W. (1977) 'Feeding behaviour of chimpanzees in Gombe National Park, Tanzania', in *Primate Ecology: Studies of Feeding and Ranging Behaviour in Lemurs, Monkeys and Apes* (Clutton-Brock, T.H., ed.) p 503–38, Academic Press, London

Yerkes, R.M. (1943) *Chimpanzees: A Laboratory Colony*, Yale Univ. Press, New Haven

Chapter 7. Reproduction and Infant Development

Antonius, J.I., Ferrier, S.A. and Dillingham, L.A. (1971) 'Pulmonary embolus and testicular atrophy in a gorilla' *Folia Primatol.* 15: p 277–92

Asano, M. (1967) 'A note on the birth and rearing of an orang-utan (*Pongo pygmaeus*) at Tama Zoo, Tokyo' *Int. Zoo Yrbk.* 7: p 95–6

Beach, F.A. (1976) 'Attractivity, perceptivity and receptivity in female mammals' *Horm. Behav.* 7: p 105–38

Brandt, E.M. and Mitchell, G. (1971) 'Parturition in primates: behavior related to birth' in *Primate Behavior Vol. 2* (L.A. Rosenblum ed.) p 177–223, Academic Press, New York

Carmichael, L., Kraus, M.B. and Reed, T.H. (1962) 'The Washington National Zoological Park gorilla infant, Tamoko' *Int. Zoo Yrbk.* 3: p 88–93

Carter, F.S. (1973) 'Comparison of baby gorillas and human infants at birth and during the post-natal period' *Ann. Rep. Jersey Wildl. Pres. Trust* 10: p 29–33

Chaffee, P.S. (1967) 'A note on the breeding of orang-utans (*Pongo pygmaeus*) at Fresno Zoo' *Int. Zoo. Yrbk.* 7: p 94–5

Coffee, P.F. (1975) 'Sexual cyclicity in captive orang-utans, *Pongo pygmaeus*, with some notes on sexual behaviour' *Ann. Rep. Jersey Wildl. Pres. Trust* 12: p 54–5

Coffey, P.F. and Pook, J. (1974) 'Breeding, hand-rearing and development of the third lowland gorilla at Jersey Zoological Park' *Ann. Rep. Jersey Wildl. Pres. Trust* 11: p 45–52

Collins, D.C., Graham, C.E. and Preedy, J.R.K. (1975) 'Identification and measurement of urinary oestrone, oestradiol 17β, oestriol, pregnanediol and androsterone during the menstrual cycle of the orang-utan', *Endocrinology* 96: p 93–101

Dixson, A.F., Moore, H.D.M. and Holt, W.V. (1980) 'Testicular atrophy in captive gorillas (*Gorilla g. gorilla*)', *J. Zool. Lond.* 191: p 315–22

Dixson, A.F., Everitt, B.J., Herbert, J., Rugman, S.M. and Scruton, D.M. (1973) 'Hormonal and other determinants of sexual attractiveness and receptivity in rhesus and talapoin monkeys', in *Primate Reproductive Behavior* (Phoenix, C.H., ed.) p 36–63, Karger, Basel.

Eaton, G.G. (1973) 'Social and endocrine determinants of sexual behavior in simian and prosimian females' in *Primate Reproductive Behavior* (Phoenix, C.H. ed.) p 20–35, Karger, Basel

Everitt, B.J., Herbert J. and Hamer, J.D. (1972) 'Sexual receptivity of bilaterally adrenalectomized female rhesus monkeys', *Physiol. Behav.* 8: p 409–15

Fisher, L.E. (1972) 'The birth of a lowland gorilla at the Lincoln Park Zoo, Chicago', *Int. Zoo Yrbk.* 12: p 106–8

Fontaine, P.A. (1968) 'Birth of four species of apes at Dallas Zoo', *Int. Zoo Yrbk.* Vol. 8: p 115–18

Frueh, R.J. (1968) 'A captive-born gorilla at St. Louis Zoo', *Int. Zoo Yrbk.* 8: p 128–31

Gibson, J.R. and McKeown, T. (1950) 'Observations on all births (23,970) in Birmingham (1947) I. Duration of gestation', *Brit. J. Soc. Med.* 4: p 221–33

Goldfoot, D.A., Kravetz, M.A., Goy, R.W. and Freeman, S.K. (1976) 'Lack of effect of vaginal lavages and aliphatic acids on ejaculatory responses in rhesus monkeys', *Horm. Behav.* 7: p 1–27

Graham, C.E. (1970) 'Reproductive physiology of the chimpanzee' in *The Chimpanzee* Vol 3 (Bourne, G.H., ed.), p 183–220, Karger, Basel

Graham, C.E., Collins, D.C., Robinson, H. and Preedy, J.R.K. (1972) 'Urinary levels of estrogens and pregnanediol and plasma levels of progesterone during the menstrual cycle of the chimpanzee: relationship to the sexual swelling', *Endocrinology* 91: p 13–24

Graham, C.E., Keeling, M., Chapman, C., Cummins, L.B. and Haynie, J. (1973) 'Method of endoscopy in the chimpanzee: relations of ovarian anatomy, endometrial histology and sexual swelling', *Amer. J. Phys. Anthrop.* 38: p 211–16

Graham-Jones, O. and Hill, W.C.O. (1962) 'Pregnancy and parturition in a Bornean orang', *Proc. Zool. Soc. London.* 139: p 503–10

Harcourt, A.H. and Stewart, K.J. (1978) 'Sexual behaviour of wild mountain gorillas', in *Recent Advances in Primatology* Vol.I *Behaviour* (Chivers, D.J. and Herbert, J., eds.) p 611–12, Academic Press, London

Harding, C.J., Danford, D. and Skeldon, P.C. (1969) 'Notes on the successful breeding by incompatible gorillas at Toledo Zoo', *Int. Zoo. Yrbk.* 9: p 84–8

Harlow, H.F. (1971) *Learning to Love*, Ballantine, New York

Heinricks, W.L. and Dillingham, A. (1970) 'Bornean orang-utan twins born in captivity', *Folia Primatol.* 13: p 150–4

Herbert, J. (1968) 'Sexual preference in the rhesus monkey (*Macaca mulatta*) in the laboratory', *Anim. Behav.* 16: p 120–8

Hinde, R.A. (1971) 'Development of social behavior' in *Behavior of Non-Human Primates* Vol. 3 (Schrier, A.M. and Stollnitz, F., eds.) p 1–68, Academic Press, New York

Hobson, B.M. (1971) 'Production of gonadotrophin, oestrogen and progesterone by the primate placenta', in *Advances in Reproductive Physiology* Vol. 5 (Bishop, M.W.H., ed.) p 67–102, Logos Press, London

Hopper, B.R., Tullner, W.W. and Gray, C.W. (1968) 'Urinary oestrogen excretion during pregnancy in a gorilla *(Gorilla gorilla)*', *Proc. Soc. Symp. Biol. Med.* 29: p 213–14

Hughes, J. and Redshaw, M. (1973) 'The psychological development of two baby gorillas: a preliminary report', *Ann. Rep. Jersey Wildl. Pres Trust* 10: p 34–6

Jantschke, F. (1972) *Orang-utans in Zoologischen Gärten*, Piper, Munich

Kagawa, M. and Kagawa K. (1972) 'Breeding a lowland gorilla at Ritsurin Park Zoo, Takamatsu,' *Int. Zoo Yrbk.* 12: p 105–6

Kirchshofer, R. (1970) 'Gorillazucht in Zoologischen Gärten und Forschungssta-tioneen,' *Der. Zool. Gärten* 38: p 73–96

Kirchshofer, R. Fradrich, H., Podolczak, D. and Podolczak, C. (1967) 'An account of the physical and behavioural development of the hand-reared gorilla infant, *Gorilla g. gorilla*, born at Frankfurt Zoo,' *Int. Zoo Yrbk.* 7: p 108–13

Kirchshofer, R., Weisse, K., Berenz, K., Klose H. and Klose I. (1968) 'A preliminary account of the physical and behavioural development during the first 10 weeks of the hand-reared gorilla twins born at Frankfurt Zoo', *Int. Zoo Yrbk.* 8: p 121–8

Knoblock, H. and Pasamanick, B. (1959) 'The development of adaptive behavior in a gorilla', *J. Comp. Physiol. Pyschol.* 52: p 699–703

Koch, W. (1937) 'Bericht uber das Ergebnis der Obduktion des gorilla "Bobby" des Zoologischen Gärtens zu Berlin. Ein Beitrag zur vergleichenden Konstitutions-pathologie', *Veroff Konst.-Wehr. Path.* 9: p 1–36

Lang, E.M. (1959) 'The birth of a gorilla at Basle Zoo', *Int. Zoo. Yrbk.* 1: p 3–7

Lang, E.M. (1962) '"Jambo", the second gorilla born at Basle Zoo', *Int Zoo Yrbk.* 3: p 84–8

Lawick-Goodall, J. van (1969) 'Some aspects of reproductive behaviour in a group of wild chimpanzees *(Pan troglodytes schweinfurthii)* at the Gombe Stream Reserve, Tanzania, East Africa', *J. Reprod. Fertil. Supp.* 16: p 353–5

Le Boeuf, B.J. (1967) 'Interindividual associations in dogs', *Behaviour* 29: p 268–95

Lindburg, D.G. and Hazell, L.D. (1972) 'Licking of the neonate and duration of labor in great apes and man,' *Amer. Anthrop.* 74: p 318–25

Mackinnon, J. (1974) *In Search of the Red Ape*, Collins, London

Mallinson, J.J.C., Coffey, P. and Usher-Smith, J. (1976) 'Breeding and hand rearing lowland gorillas, *Gorilla g. gorilla*, at the Jersey Zoo', *Int. Zoo Yrbk* 16: p 189–94

Marshall, W.A. (1968) *Development of the Brain*, Oliver and Boyd, Edinburgh

Martin, R.D., Seaton B., and Lusty J. (1975) 'Application of urinary hormone determinations in the management of gorillas', *Ann. Rep. Jersey Wildl. Pres. Trust* 12: p 61–70

McCormack, S.A. (1971) 'Plasma testosterone concentration and binding in the chimpanzee: effect of age', *Endocrinology* 89: p 1171–7

McKenney, F.D., Traum, J. and Bonestell, A.E. (1944) 'Acute coccidiomycosis in a gorilla *(Gorilla beringei)* with anatomical notes', *J. Am. Vet. Med. Assoc.* 104: p 136–40

Michael, R.P. and Keverne, E.B. (1970) 'Primate sex pheromones of vaginal origin' *Nature* 225: p 84–5

Nadler, R.D. (1974) 'Periparturitional behaviour of a primiparous lowland gorilla', *Primates* 15: p 55–73

Nadler, R.D. (1974) 'Determinants of variability in maternal behavior of captive female gorillas' in *Proc. 5th Int. Cong. Primatol.* (Kondo, S. and Kawai, M., eds.) p 207–16, Japan Science Press, Tokyo

Nadler, R.D. (1975) 'Second gorilla birth at Yerkes Primate Research Center', *Int. Zoo Yrbk.* 15: p 134–7

Nadler, R.D. (1975) 'Cyclicity in tumescence of the perineal labia of female lowland gorillas', *Anat. Rec.* 181: p 791–7

Nadler, R.D. (1975) 'Sexual cyclicity in captive lowland gorillas', *Science* 189: p 813–14

Nadler, R.D. (1976) 'Sexual behavior of captive lowland gorillas', *Arch. Sex. Behav.* 5: p 487–502

Nadler, R.D. (1978) 'Sexual behaviour of orang-utans in the laboratory', in *Recent Advances in Primatology* Vol 1. *Behaviour* (Chivers D.J. and Herbert, J., eds.) p 607–8 Academic Press, London

Nadler, R.D., Graham, C.E., Collins, D.C. and Gould, K.G. (1979) 'Plasma gonadotropins, prolactin, gonadal steroids and genital swelling during the menstrual cycle of lowland gorillas', *Endocrinology* 105: p 290–6

Nadler, R.D. and Jones, M.L. (1974) 'Breeding the gorilla in captivity', *Int. Zoo News* 22: p 21–7

Nissen, H.W. and Yerkes, R.M. (1943) 'Reproduction in the chimpanzee: report on forty-nine births', *Anat. Rec.* 86: p 567–78

Noback, C.R. (1939) 'The changes in the vaginal smears and associated cyclic phenomena in the lowland gorilla (*Gorilla g. gorilla*)', *Anat. Rec.* 73: p 209–25

Redshaw, M. (1978) 'Cognitive development in human and gorilla infants', *J. Human Evol.* 7: p 133–48

Reed, T. and Gallagher, B.F. (1963) 'Gorilla birth at National Zoological Park, Washington', *Der Zool. Garten* 27: p 279–92

Rowell, T.E. (1970) 'Baboon menstrual cycles affected by social environment', *J. Reprod. Fertil.* 21: p 133–41

Rumbaugh, D.M. (1967) '"Alvila" – San Diego Zoo's captive-born gorilla' *Int. Zoo Yrbk.* 7: p 98–107

Savage, S. and Bateman, R. (1978) 'Sexual morphology and behaviour in *Pan paniscus*' in *Recent Advances in Primatology* Vol. 1 *Behaviour* (Chivers, D.J., and Herbert, J. eds.) p 613–16 Academic Press, London

Steiner, P.E., Rasmussen, T.B., Fisher, L.E. (1955) 'Neuropathy, Cardiopathy, Hemosiderosis and Testicular Atrophy in *Gorilla gorilla*', *Arch. Path.* 59: p 5–25

Stewart, K.J. (1977) 'The birth of a wild mountain gorilla (*Gorilla gorilla beringei*)', *Primates* 18: p 965–76

Tanner, J.M. (1962) *Growth at Adolescence*, Blackwell Sci. Publ., London

Thomas, W.D. (1958) 'Observation on the breeding in captivity of a pair of lowland gorillas', *Zoologica* 43: p 95–104

Tijskens, J. (1971) 'The oestrous cycle and gestation period of the mountain gorilla', *Int. Zoo Yrbk.* 11: p 181–3

Yerkes, R.M. and Elder, J.H. (1936) 'Oestrus receptivity and mating in chimpanzee', *Comp. Psychol. Monogr.* 13: p 1–39

Yerkes, R.M. and Elder, J.H. (1936) 'The sexual and reproductive cycles of chimpanzee', *Proc. Nat. Acad. Sci.* 22: p 276–83

Young, W.C. and Orbison, W.D. (1944) 'Changes in selected features of behavior in pairs of oppositely-sexed chimpanzees during the sexual cycle and after ovariectomy', *J. Comp. Psychol.* 37: p 107–43

Young, W.C. and Yerkes, R.M. (1943) 'Factors affecting the reproductive cycle in the chimpanzee: the period of adolescent sterility and related problems', *Edocrinology* 33: p 121–54

Younglai, E.V., Collins, D.C. and Graham, C.E. (1977) 'Progesterone metabolism in female gorilla', *J. Endocr.* 75: p 439–40

Chapter 8. Conservation or Extinction?

Akeley, M.L.J. (1931) *Carl Akeley's Africa*, Gollancz, London

Akeley, M.L.J. (1950) *Congo Eden*, Dodd, Mead, New York

Blancou, L. (1950) 'The lowland gorilla', *Animal Kingdom* 58: p 162–9

Blancou, L. (1961) 'Destruction and protection of the wildlife in French Equatorial and French West Africa Part 5: Primates', *African Wildlife* 15: p 29–34

Caldwell, K. (1934) 'International conference for the protection of fauna and flora in Africa', *J. Soc. Pres. Fauna Empire* Part 22

Cousins, D. (1978) 'Gorillas – a survey', *Oryx* 14: p 254–8 and 374–6

Critchley, W. (Unpublished manuscript 1968) 'Report of the Takamanda gorilla survey'

Gartlan, J.S. (1974) 'The African forests and problems of conservation', in *Proc. 5th Int. Congr. Primatol.* (Kondo S. and Kawai, M., eds.) p 509–24, Japan Science Press, Tokyo

Gartlan, J.S. (Unpublished document and personal communication 1974) Memorandum on strategy for an I.U.C.N./W.W.F. gorilla conservation project 1. The western gorilla

Goodall, A.G. and Groves, C.P. (1977) 'The conservation of eastern gorillas', in *Primate Conservation* (Prince Rainier III of Monaco and G.H. Bourne eds.) p 599–637, Academic Press, New York

Goodwin, H.A. and Holloway, C.W. (eds.) (1972) *The Red Data Book* Vol. 1 *Mammalia* I.U.C.N.

Harcourt, A.H. (1976) 'Virunga gorillas – the case against translocations', *Oryx* 13: p 469–72

Harcourt, A.H. (1979) 'F.P.S. Mountain Gorilla Project: progress report 2', *Oryx* 15: p 114–15

Harcourt, A.H. (1980) 'The Mountain Gorilla Project: a two-year progress report', *Primate Eye* 14: p 13–18

Harcourt, A.H. and Curry-Lindahl, K. (1978) 'The F.P.S. Mountain Gorilla Project – a report from Rwanda', *Oryx* 14: p 316–324

Harrison, B. (ed.) (1971) *Primates in Medicine*, Vol. 5: *Conservation of Non-Human Primates in 1970*, Karger, Basel

Mackinnon, J. (1976) 'Mountain gorillas and bonobos', *Oryx* 13: p 372–82

Martin, R. D. (ed.) (1975) *Breeding Endangered species in Captivity*, Academic Press, London

Verschuren, J. (1975) 'Wildlife in Zaire', *Oryx* 13: p 25–33 and 149–63

Webb, J. (Unpublished manuscript) 'Factors in the distribution of *Gorilla gorilla* – a review of the literature and preliminary fieldwork in the Cameroun'

Index

199